Managing Post-Quantum Cryptography

Securing the Enterprise Before the Machines Catch Up

Claude Louis-Charles, Phd

i

Cybersoft Publishing LLC

Fort Washington, MD 20744

Drclaude.net

First Edition March 2026

Table of Contents

Foreword

Every encrypted message your organization sent today — every financial transaction, every patient record transfer, every classified communication — rests on a mathematical problem that a sufficiently powerful quantum computer will solve. Not theoretically. Not in some distant academic scenario. The algorithms are already published. The standards to replace the vulnerable ones have already been finalized. The only variable is the timeline, and the organizations that wait for certainty on that timeline will find themselves executing a migration under crisis conditions that should have been a planned transition.

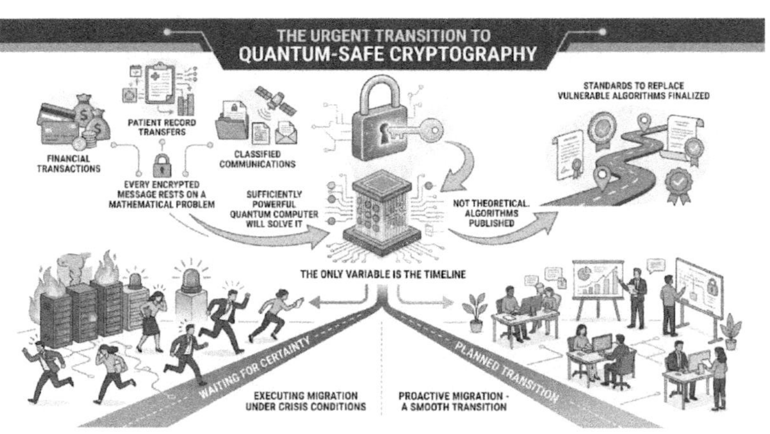

I wrote this book because I watched the Y2K remediation unfold from inside federal technology programs, and the parallels to the post-quantum migration are uncomfortably precise — with one critical difference.

When the calendar flipped in 2000, every organization knew exactly when the deadline arrived. The quantum threat carries no fixed date. That ambiguity is not a reason for patience. This is why urgency matters more: there is no external forcing function that will compel your organization to act until the damage is already irreversible.

The technical community has done its part. The National Institute of Standards and Technology has selected and published the algorithms. Hardware and software vendors have begun integrating them. What remains is the enterprise execution — the cryptographic inventory, the dependency mapping, the hybrid deployment strategy, the governance model that keeps the migration moving through budget cycles and leadership changes. That execution challenge is where most organizations will struggle, and it is precisely where this book focuses.

This is not a book about quantum physics. You will not need a background in number theory or lattice mathematics to use it. It is a book about managing a technology transition that touches every system in your enterprise that relies on confidentiality, integrity, or authentication — which is to say, nearly everything. Each chapter is designed to give you the vocabulary to engage your technical teams, the frameworks to sequence your decisions, and the governance structures to sustain the migration over the years it realistically will require.

The window for acting ahead of this threat is open now, but it is not indefinite. Adversaries are already harvesting encrypted traffic with the explicit intent of decrypting it later. Every month your organization delays the cryptographic inventory is another month of data that may be exposed retroactively. This book exists to help you close that window on your terms — methodically, confidently, and before the machines catch up.

Introduction

This book exists at the intersection of two realities that most enterprise leaders have not yet reconciled. The first reality is that every cryptographic system protecting your organization today was designed to resist classical computers, and a sufficiently capable quantum computer will defeat the mathematical assumptions on which those systems depend. The second reality is that migrating an enterprise cryptographic infrastructure is a multiyear effort that touches every application, certificate, key store, and protocol in the environment. The nine chapters ahead are designed to close the gap between those two realities by giving you a structured path from awareness to execution.

The opening three chapters build the technical literacy you will need to lead the transition without becoming a cryptographer yourself. You will learn what quantum computing actually threatens — and what it does not — followed by an accessible treatment of the new algorithmic families that replace the vulnerable ones, and a detailed walkthrough of the standards that the National Institute of Standards and Technology has finalized. These chapters are written for the leader who needs to understand the material well enough to ask the right questions, allocate resources accurately, and evaluate vendor claims with confidence.

Chapters four through six move into the operational core of the migration. The cryptographic census — finding every key, certificate, and algorithm your organization depends on — is treated as its own chapter because it is the single most underestimated phase of the transition. From there, you will work through the hybrid deployment strategy that allows classical and post-quantum algorithms to operate in parallel during the migration window, followed by the crypto-agility architecture that ensures your enterprise never faces this kind of forced migration again.

The final three chapters address execution at scale. You will find a detailed migration playbook with sequencing guidance, risk-tiered prioritization, and integration checkpoints. The closing chapter establishes the governance model, workforce development strategy, and long-term monitoring framework that turns a one-time migration into an ongoing practice of cryptographic resilience.

Each chapter is self-contained enough to serve as a reference when a specific question arises — your team can pull the cryptographic census chapter during inventory planning or the hybrid deployment chapter when evaluating vendor integration approaches. For your first reading, the sequential path is recommended because each chapter builds on the decision context assumed by the next. When technical depth exceeds what a particular reader needs, the practical takeaway is stated plainly before moving forward, so no one loses the thread.

1 The Clock in the Cipher: Why the Encryption Protecting Your Enterprise Has an Expiration Date

Scenario: A federal healthcare agency stores patient records encrypted with RSA-2048. The security team last reviewed the cryptographic configuration three years ago, when the auditors signed off without a comment. The CISO knows quantum computing exists, but has filed it mentally under "future problem." Then a brief appears on her desk: CISA has quietly notified several sector partners that a foreign intelligence service is systematically archiving encrypted inter-agency traffic. The traffic is unreadable today. The classification on the brief is a reminder that it may not stay that way. The CISO calls her architecture lead. How long have we had this configuration? The answer is nine years. How quickly can we swap it out? Silence follows.

This chapter is about that silence — what causes it, what it costs, and how to replace it with a plan—every encryption scheme your enterprise depends on rests on a mathematical assumption about difficulty. The difficulty is real, but it is not permanent. Quantum computing does not break cryptography by brute force in the classical sense; it exploits structural weaknesses in the specific mathematical problems on which asymmetric cryptography is based. Understanding that distinction is the first step toward making decisions that protect your organization before the window closes.

1.1 The Mathematical Bargain Underlying Modern Encryption

Every security decision your team makes on certificates, TLS configurations, VPN authentication, and digital signatures is underwritten by a single principle: certain mathematical operations are easy to perform in one direction and computationally infeasible to reverse. This is not a metaphor. It is the literal technical foundation for billions of encrypted transactions executed every day. Managers who understand this principle make better procurement decisions, ask sharper questions of their vendors, and recognize why the quantum threat is structurally different from every other cryptographic risk they have managed before.

1.1.1 Hardness Assumptions That Classical Security Depends On

Modern public-key cryptography relies on three primary hardness assumptions. The first is integer factorization: given a large number that is the product of two large primes, finding those primes is computationally intractable for any known classical algorithm when the number is large enough. RSA, which underpins a significant share of enterprise certificate infrastructure, HTTPS, and secure email, is built on this assumption. The second is the discrete logarithm problem: given a group element and a generator, finding the exponent that relates them is hard in certain mathematical groups. Diffie-Hellman key exchange and its variants depend on this. The third is the elliptic curve discrete logarithm problem: the same fundamental challenge, but defined over the geometric structure of elliptic curves. Elliptic Curve Cryptography, or ECC, exploits this, which is why it achieves strong security with shorter keys than RSA.

These assumptions have been tested continuously since the 1970s. The mathematical community has attempted to solve them using classical algorithms for decades, and the best-known approaches remain exponential in the size of the key. That exponential growth is what makes a 2048-bit RSA key feel safe: the number of operations required to factor it classically exceeds what any realistic compute cluster could complete within a human lifetime. The assumption is not that the problem is impossible. The assumption is that it is hard enough, for long enough, given the tools available. That last clause is where quantum computing changes the calculation.

1.1.2 Why "Computationally Infeasible" Is Not the Same as "Impossible."

Security practitioners often speak of "unbreakable" encryption in casual conversation, but no cryptographic scheme is unconditionally secure in the information-theoretic sense. What cryptography provides is computational security: breaking the scheme requires more computational resources than are practically available within the useful lifetime of the protected data. When a scheme is described as secure, what that really means is that no known algorithm can break it efficiently given current and anticipated computing resources. Both halves of that sentence matter. "Known algorithm" is a research boundary, not a permanent wall. "Current and anticipated computing resources" is a forecast, not a fact.

The history of cryptography includes repeated examples of hardness assumptions that were believed to be robust but later proved fragile. MD5 and SHA-1 were widely deployed before collision attacks rendered them unsuitable for security-critical applications. Export-grade cipher suites were mandated by policy and then weaponized in downgrade attacks a generation later. The lesson is not that cryptography cannot be trusted, but that trust must be calibrated to the state of the art in attack capability — and that state of the art changes. The quantum era is not a break from this pattern. It is an acceleration of it, driven by a fundamentally different computational model that attacks problems classical computers cannot.

Diagram 1.1 – The Hardness Assumption Lifecycle

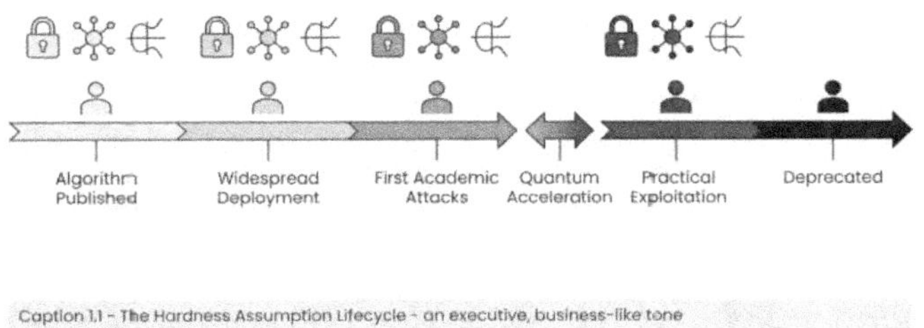

Caption 1.1 - The Hardness Assumption Lifecycle - an executive, business-like tone

1.1.3 The Asymmetric Key Paradigm and Its Structural Fragility

The design of asymmetric cryptography is elegant in its economics: a key pair is generated, with the public key freely distributed while the private key remains secret. Anything encrypted with the public key can only be decrypted with the private key. Anything signed with the private key can be verified by anyone holding the public key. This architecture powers certificate authorities, TLS handshakes, code signing pipelines, and every form of identity verification that does not require pre-shared secrets. It is also architecturally concentrated: if the hardness assumption underlying the key pair collapses, every use of that key pair — and every key pair in the same algorithm family — loses its security property simultaneously.

This concentration is what makes the quantum threat different from, say, a vulnerability in a single product or a misconfiguration in a specific server. A classical vulnerability typically affects a bounded scope: a version, a vendor, a specific deployment. A break in the hardness assumption underlying RSA or ECC does not respect those boundaries. It affects every certificate ever issued, every key ever generated, every encrypted archive ever stored, regardless of who issued it, what product created it, or how carefully it was managed. Your enterprise does not have a single RSA deployment; it has hundreds or thousands spanning applications, network equipment, development pipelines, HSMs, identity platforms, and vendor integrations. That is the scope of the migration.

1.1.4 Symmetric Ciphers, Key Exchange, and the Threat Surface They Share

Symmetric encryption — where the same key encrypts and decrypts data — is not vulnerable to quantum computing in the same catastrophic way that asymmetric cryptography is. AES-256 remains structurally sound against quantum attacks with an adequate key size, though Grover's algorithm provides a quadratic speedup in key search, effectively halving the security margin. What that means in practical terms is discussed in Chapter 2. The point is that the risk profile differs between symmetric and asymmetric schemes, and this distinction drives remediation priorities. Your most urgent exposure is in key exchange and authentication — the protocols that use RSA, DH, or ECC — not in bulk data encryption, which AES performs once a session key is established.

Key exchange is where the two worlds collide. TLS, for example, uses asymmetric cryptography to establish a session key, and then uses that session key with a symmetric cipher for the actual data transfer. An attacker who can break the asymmetric exchange can recover the session key and decrypt the traffic, even if the symmetric cipher itself is sound. This is why the migration priority is not simply "replace everything" but "replace the asymmetric layer first, then assess the symmetric layer." Understanding this distinction gives managers a basis for phased planning and for defending resource allocation to leadership.

1.2 Quantum Acceleration: From Laboratory Curiosity to Strategic Threat

The quantum computing threat to cryptography is not a theoretical speculation published in a research paper that no one reads. It is an active area of national security investment, private-sector competition, and academic advancement. The transition from "researchers are exploring this" to "adversaries are acting on it" has already occurred — not because a cryptographically relevant quantum computer exists today, but because sophisticated actors have concluded that it will exist within a strategic planning horizon and are positioning accordingly.

Diagram 1.2 – The Quantum Threat Readiness Timeline.

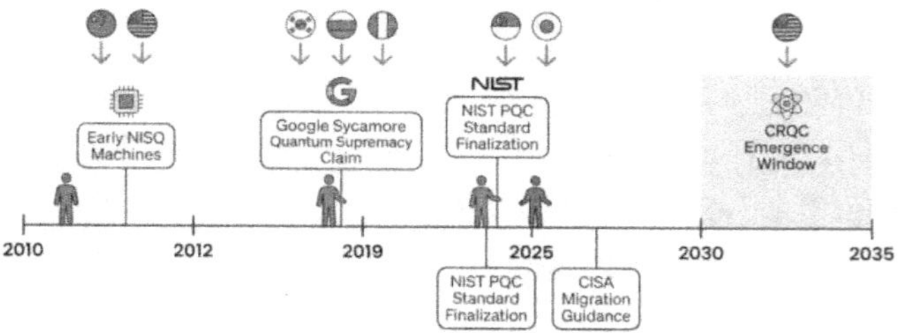

NIST PQC standardization marks a critical step for preparing organizations for future CRQC emergence.
CISA Migration Guidance provides a practical roadmap for post-quantum cryptographic readiness.

1.2.1 The Trajectory of Qubit Counts and Error-Correction Progress

Quantum hardware has advanced rapidly. Early quantum processors operated with a handful of physical qubits and error rates that made sustained computation impossible. Within a decade, leading systems have scaled to hundreds and then thousands of physical qubits, with demonstrable improvements in coherence times and gate fidelity. The critical threshold for cryptographically relevant computation requires not just raw qubit counts but logical qubits — error-corrected units that can sustain computation reliably over many operations. The overhead for error correction is substantial: each logical qubit currently requires many physical qubits. The precise ratio depends on the error rate and the correction code used. This overhead is the primary engineering barrier between today's systems and cryptographic relevance.

For a manager, the relevant question is not the exact number of qubits required to break RSA-2048 — a number that fluctuates as algorithmic improvements emerge — but rather the trajectory of engineering progress. Hardware roadmaps from major quantum computing programs show consistent qubit scaling. Error correction research is producing new codes with improved overhead efficiency. Investment in cryogenic control systems, which are a practical constraint on qubit counts, is accelerating. The pattern is not linear improvement but compounding progress across multiple enabling disciplines simultaneously. That pattern has historically preceded technological capability thresholds faster than incumbents anticipated.

1.2.2 Nation-State Investment and the Geopolitics of Quantum Supremacy

The quantum computing race is not merely a commercial competition. It is a strategic priority for multiple nation-states that have explicitly identified cryptographic capabilities as national security objectives. Governments have allocated billions in public funding to quantum research programs. Military and intelligence agencies have integrated quantum capability development into long-range planning. The concern is not symmetric: an adversary that achieves a cryptographically relevant quantum computer before defensive migration is complete can retroactively decrypt archived communications that were captured under the assumption that they would never be deciphered. That asymmetry — an offensive advantage that applies to both past and future communications — is unlike almost any other technology gap in the security domain.

The geopolitical context shapes the migration urgency in a specific way. Classified assessments from intelligence communities in multiple countries have prompted policy responses — federal mandates, migration deadlines, standards-setting processes — that are calibrated to a threat timeline that is not fully public. When federal agencies issue guidance with specific migration deadlines, those deadlines are not arbitrary. They reflect planning assumptions based on assessments that are not shared with the general public. Enterprise security leaders who treat federal guidance as bureaucratic overhead rather than mission-aligned threat intelligence are making a strategic error. The timeline embedded in those mandates represents an outside view of when the window closes.

1.2.3 Why Enterprise Security Teams Cannot Wait for Public Announcements

A common objection to post-quantum migration investment goes like this: we will act when the threat is real, and the threat is not real yet because no one has demonstrated a cryptographically relevant quantum computer. This reasoning has two structural problems. The first is that by the time a cryptographically relevant quantum computer is confirmed and publicly announced, the migration window for protecting current data will already have closed for any organization that has not migrated. The second is that the "harvest now, decrypt later" threat does not require a future quantum computer to be a present risk. It only requires an adversary to believe that a quantum computer will eventually exist and to act on that belief today by capturing and archiving encrypted traffic.

The practical implication is that the migration investment must precede the threat's confirmation by years, not months. Organizations that wait for public confirmation will be protecting only the data generated after they begin migrating, not the data already captured. For organizations that hold sensitive data with long confidentiality horizons — defense contractors, healthcare providers, financial institutions, intelligence community partners — this distinction is not academic. It is the difference between an effective defense and a migration that arrives too late to protect the data that actually matters.

1.3 Harvest Now, Decrypt Later: The Threat That Has Already Started

The most operationally significant implication of quantum computing for enterprise security leaders does not require a quantum computer to exist right now. It requires only that adversaries believe one will exist within their planning horizon — and act accordingly. The harvest now, decrypt later threat posture is the practice of capturing and storing encrypted communications and data today with the intention of decrypting them later, once the necessary computational capability is available. This threat is already active. It changes the threat model for sensitive communications that were never intended to be decrypted.

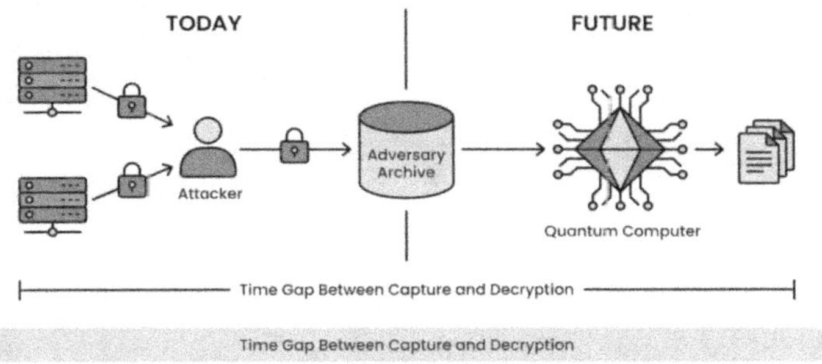

1.3.1 How Adversaries Are Stockpiling Encrypted Traffic Today

Adversary collection of encrypted network traffic is not hypothetical. It has been documented in multiple national security contexts. Intelligence assessments, Congressional testimony, and declassified reporting all point to systematic efforts by state actors to collect and store large volumes of encrypted traffic from government, defense, and critical infrastructure networks. The marginal cost of storing encrypted data has collapsed over time — storage is cheap enough that archiving traffic at scale is operationally feasible. What changes in the future is not the cost of storage but the cost of decryption. When a cryptographically relevant quantum computer becomes available, the stored archive becomes retroactively readable.

For enterprise security teams, the practical implication is that the confidentiality of data transmitted over classical cryptographic protocols today cannot be guaranteed beyond the estimated arrival of quantum computing capability. This is not a speculative risk. It is a structural feature of the threat environment that sophisticated adversaries have operationalized. The question is not whether this is happening but whether your organization holds data whose exposure window — the time between capture and the availability of a decryption capability — overlaps with the confidentiality horizon of that data.

1.3.2 Data with Long Confidentiality Horizons: Identifying Your Exposure

Not all data has the same confidentiality horizon. A payment transaction that completes and settles within days has a short horizon — even if it is intercepted and archived today, its business value expires quickly. A patent application, a diplomatic communication, a patient health record, a defense acquisition plan, or a personnel file has a horizon measured in years or decades. For that data, the harvest now threat is direct and material. If it was transmitted over a network using protocols secured by classical asymmetric cryptography, and if a sophisticated adversary collected it, then the security of that data depends on whether you migrate before a quantum computer arrives.

Identifying your long-horizon data is the first step in scoping your exposure. This is not a technical exercise alone. It requires input from legal, compliance, records management, and business leadership to determine which data categories the organization generates and transmits that carry confidentiality requirements extending beyond five years. Once those categories are identified, the question becomes: what cryptographic protocols protect the transmission and storage of that data, and how quickly could those protocols be migrated to post-quantum equivalents? This is a workflow-level impact assessment that connects cryptographic risk to business and mission outcomes.

1.3.3 Federal and Healthcare Environments as Priority Targets

Federal agencies and healthcare organizations occupy a particular position in the current threat landscape. Both sectors generate and transmit data with long confidentiality horizons as part of routine operations. Federal agencies handle classified or sensitive national security information, personnel records, law enforcement data, and infrastructure assessments — all categories with confidentiality requirements that extend for years or decades. Healthcare organizations hold protected health information, genetic data, behavioral health records, and clinical trial data — categories in which exposure can harm individuals long after the data were generated.

Both sectors are also known targets of sophisticated state actors. Federal agency networks are probed continuously. Healthcare organizations have been targeted in waves of ransomware and data exfiltration operations. The combination of high-value, long-horizon data and established adversary interest makes these sectors the highest-priority targets for harvest-now collection. Federal guidance specifically addresses this: NSA, CISA, and OMB have issued migration timelines and requirements that reflect the assessment that federal agencies face an elevated threat. Healthcare organizations operating under HIPAA have compliance obligations that are beginning to incorporate post-quantum considerations. The regulatory and threat environments are converging around the same urgency.

Diagram 1.4 – Data Confidentiality Horizon vs. Migration Urgency

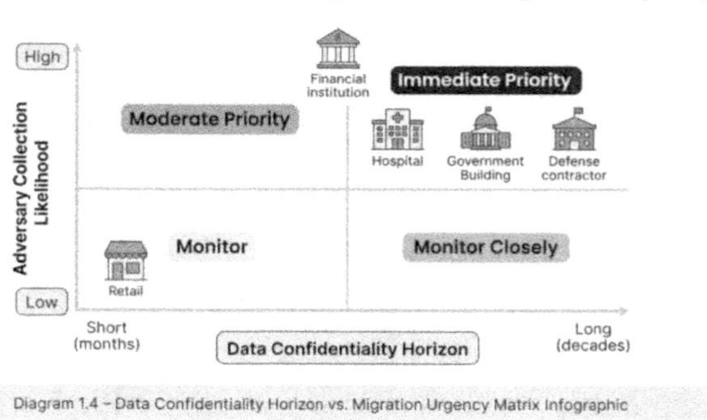

Diagram 1.4 – Data Confidentiality Horizon vs. Migration Urgency Matrix Infographic

1.4 The Migration Complexity That Makes This Different from Prior Upgrades

Every major platform team has managed a cryptographic migration at some point: a certificate renewal, a TLS version upgrade, a hash algorithm deprecation. Those migrations were hard. Post-quantum migration is an order of magnitude harder. The scope is broader, the dependencies are deeper, the testing requirements are more demanding, and the organizational change management challenge is more complex. Understanding why this migration is different is necessary to scope it correctly and set realistic expectations with leadership.

1.4.1 Cryptographic Surface Area in a Modern Enterprise

The cryptographic surface area of a modern enterprise extends far beyond the places where IT teams typically think about cryptography. The obvious areas are well-known: TLS certificates on web servers, VPN configurations, email security, and identity platforms. The less obvious areas are where migrations stall. Custom applications that include cryptographic libraries — sometimes multiple versions, sometimes embedded in vendor components — constitute a significant, often unmapped population of cryptographic dependencies. Network equipment: switches, routers, load balancers, and out-of-band management interfaces all implement cryptographic protocols in firmware that may or may not be updatable to support post-quantum algorithms.

IoT and operational technology devices are particularly challenging. Many of these devices have fixed firmware, limited memory, and no mechanism for algorithm updates. If they participate in cryptographically secured communications — and most do, for authentication, firmware integrity, or data transmission — they represent a long-tail problem in the migration. HSMs, which are specifically designed to be cryptographic hardware roots of trust, require vendor firmware updates that may not be available on existing hardware, triggering hardware replacement cycles. PKI infrastructure, which issues and manages certificates for the entire enterprise, must itself be migrated before the assets it serves can be migrated.

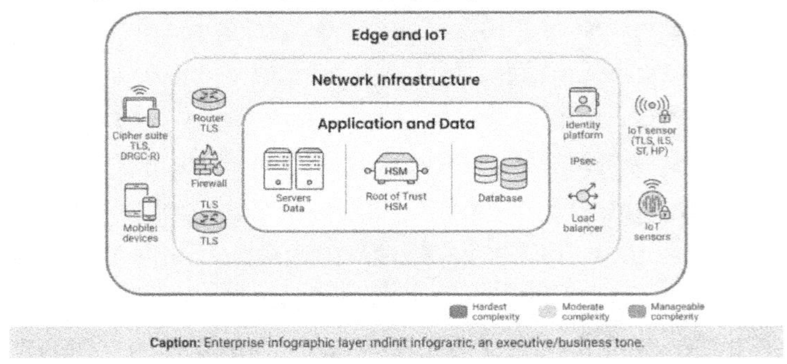

Caption: Enterprise infographic layer indinit infogramic, an executive/business tone.

1.4.2 Why Swapping Algorithms Is Not Like Patching a Vulnerability

A patch addresses a specific defect in a specific component. It is applied, tested, and closed. An algorithm migration affects the interface between components — the protocols they speak, the key formats they accept, and the certificate structures they validate. Changing an algorithm, therefore, requires coordinated changes across all endpoints participating in a given protocol exchange. If you migrate your web servers to a post-quantum key exchange algorithm but your load balancers do not support it, the connection negotiation fails. If you issue post-quantum certificates from your CA but your application servers cannot parse them, authentication breaks. Every hop in a communication path must be updated in a coordinated sequence, and the sequence must be tested before deployment.

This coordination requirement is what makes algorithm migration fundamentally different from vulnerability patching. It is not a product-level change but an ecosystem-level change. It requires vendor readiness across the supply chain, not just internal engineering capacity. It requires interoperability testing of protocols across heterogeneous environments. It requires a sequenced rollout plan that migrates foundations — certificate infrastructure, key management systems — before dependents. And it requires rollback planning at every stage, because a failed migration in a critical authentication path can cause a service outage that is indistinguishable from a security incident.

1.4.3 The Organizational Inertia Problem in Large-Scale Crypto Transitions

Even when the technical plan is solid, large-scale cryptographic migrations fail on organizational grounds. Cryptography is infrastructure — it is invisible when it works and catastrophic when it does not. That invisibility creates organizational inertia. Application teams have no incentive to prioritize a migration that adds latency, requires testing, and offers no visible user-facing improvement. Procurement teams are not equipped to evaluate vendor post-quantum readiness without guidance. Audit teams may not yet have a framework for evaluating compliance with migration requirements. And executive leadership, absent a specific incident, may not see the urgency of a multiyear investment in something that is not broken today.

Breaking organizational inertia requires two things: a clear risk narrative that connects cryptographic exposure to business and mission outcomes, and a program structure that distributes migration responsibility across organizational units with accountable owners. Neither of these is purely technical. The risk narrative must be translated into the language of business continuity, regulatory compliance, and reputational risk. The program structure must align with the organization's existing governance model — a standalone security initiative with no executive sponsorship will stall. The chapters ahead address both the technical and organizational dimensions in detail. This chapter establishes why the urgency is real before those specifics matter.

1.5 Reframing the Risk Conversation for Leadership

Cryptographic risk is difficult to communicate at the executive level because it is abstract, technical, and probabilistic in nature. The risk is not a specific incident that has already happened; it is a class of incidents that will happen if action is not taken within a window that is itself uncertain. This communication challenge is real and not unique to post-quantum risk — it is the same challenge that security leaders face when presenting any long-horizon strategic risk to leadership that is focused on near-term operational performance. The techniques for navigating it are available, but they require deliberate reframing.

Diagram 1.6 – Translating Cryptographic Risk to Business Language

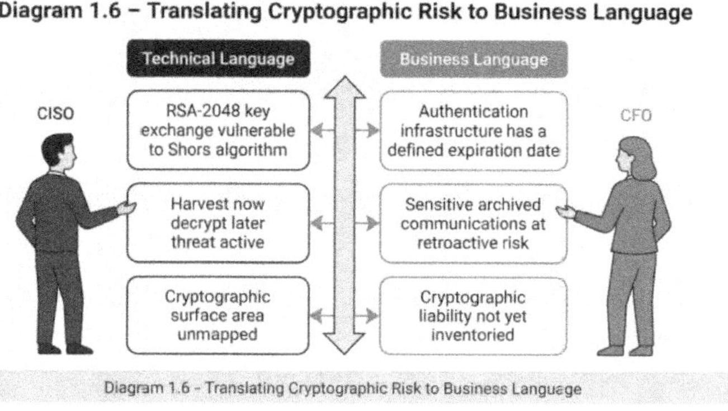

Diagram 1.6 – Translating Cryptographic Risk to Business Language

1.5.1 Translating Cryptographic Risk into Business and Mission Impact

The most effective translation of cryptographic risk into executive language is mission-aligned. For a defense contractor, the mission impact of a post-quantum breach is the exposure of acquisition plans, technical specifications, or personnel data to a foreign adversary. For a hospital system, the mission impact is the retroactive exposure of patient records to actors who could use that data for identity fraud, insurance manipulation, or political leverage. For a financial institution, the mission impact includes authentication failures that allow fraudulent transactions, certificate forgery that enables man-in-the-middle attacks on client communications, and regulatory penalties for failure to maintain adequate security controls.

Each of these framings connects cryptographic risk to something the organization already cares about and has frameworks for managing. Business continuity risk, regulatory compliance risk, and reputational risk are categories that boards, executives, and audit committees regularly address. Post-quantum migration belongs in each of those categories. The manager's role is to build that translation bridge explicitly — not to expect that the technical detail will be persuasive on its own, and not to present the risk in a way that induces paralysis rather than action.

1.5.2 What "Cryptographically Relevant Quantum Computer" Actually Means for Timelines

The term "cryptographically relevant quantum computer" — often abbreviated CRQC — refers to a quantum computer with sufficient scale, error correction, and operational reliability to execute Shor's algorithm on the key sizes used in real-world cryptographic deployments. A CRQC for RSA-2048 would require millions of physical qubits with current error correction approaches, or substantially fewer with future advances in error-correcting codes. Today's machines are orders of magnitude below that threshold. But the relevant planning question is not where the hardware is today; it is what trajectory of progress to assume for a five-to-fifteen-year planning horizon.

Responsible estimates from serious technical sources project CRQC capability arriving somewhere in the 2030s, with high uncertainty in both directions. Some scenarios place it earlier; some place it later. The planning implication is that organizations with long migration lead times — those with complex infrastructure, many vendor dependencies, or limited internal cryptographic engineering capacity — should begin their migrations now, not wait until the CRQC timeline becomes clearer. If a migration requires five to seven years for a large enterprise, and a CRQC arrives in 2032, an organization that begins in 2030 will not finish in time. Operational clarity on this point is what converts the abstract risk into a concrete schedule.

1.5.3 Setting Executive Expectations Without Inducing Paralysis

Communicating quantum risk to leadership without causing one of two failure modes — dismissal ("this is too far in the future to act on now") or panic ("everything we have is already broken") — requires calibrated framing. The dismissal failure mode is addressed by grounding the urgency in the harvest now threat, which is present-tense, and in regulatory migration deadlines, which have specific dates. The panic failure mode is addressed by presenting the risk alongside a concrete phased action plan that makes the migration tractable and shows measurable progress within each budget cycle.

The migration is large, but it is not chaotic. It has a defined scope — the organization's cryptographic surface area — that can be inventoried and prioritized. It has a defined solution — NIST-standardized post-quantum algorithms — that has been through rigorous public evaluation. It has a defined process — discovery, prioritization, piloting, phased deployment — that can be managed like any other large infrastructure program. Presenting it this way to leadership does not minimize the risk; it demonstrates that the organization is approaching a serious threat with appropriate seriousness and professional competence. That combination of honesty and operational clarity is what earns the sustained executive sponsorship that this migration requires.

1.6 Manager's Checklist: Chapter 1

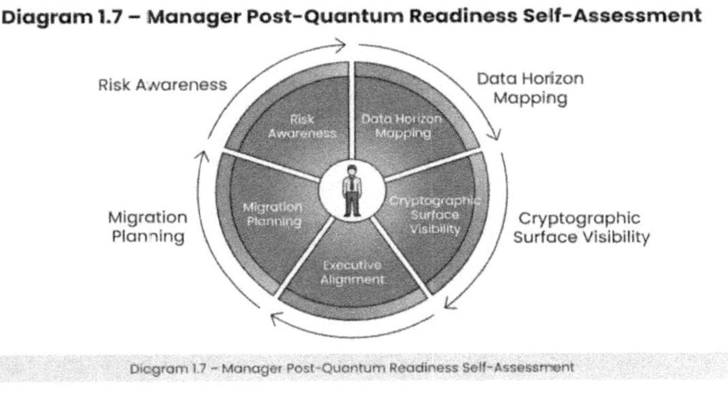

Diagram 1.7 – Manager Post-Quantum Readiness Self-Assessment

Diagram 1.7 – Manager Post-Quantum Readiness Self-Assessment

- Confirm whether your security team has completed any cryptographic asset inventory in the last 18 months. If not, initiate one.
- Identify the three to five categories of sensitive data your organization holds with confidentiality horizons exceeding five years.
- Request from your network and security teams a list of the external communications protocols actively in use: TLS versions, key exchange algorithms, certificate signature algorithms.

- Determine whether your organization has received any CISA, NSA, or sector-specific guidance on post-quantum migration and verify whether it has been formally reviewed by leadership.
- Assess whether executive leadership can articulate the harvest now threat and its relevance to your data environment. If not, prepare a briefing.
- Confirm whether your largest vendors — network hardware, PKI, identity platforms, cloud providers — have published post-quantum roadmaps. If not, prepare questions for the next procurement or renewal cycle.
- Identify who in your organization owns cryptographic policy decisions and who would own a migration program. If those roles are vacant or unclear, escalate.
- Determine whether your organization has a documented cryptographic policy that specifies algorithm standards and review cycles. If the policy is more than three years old, flag it for update.

Diagram 1.8 – The Organizational Inertia Problem in Crypto Migration

Caption. The organizantin tinic management ootiluate arrrav prograsts anncnnls in crypto migration

1.7 The Ground Has Already Shifted

The encryption protecting your enterprise was designed for a world in which certain mathematical problems could not be solved efficiently. That world is ending — not in a single moment, but through a trajectory of hardware and algorithmic advances that is already underway. The harvest now threat means that the migration does not start when a quantum computer is confirmed; it started the day your organization's sensitive communications became worth archiving.

The migration ahead is large and complex, but it is not unprecedented. It has a defined scope, a maturing solution set anchored in NIST-standardized algorithms, and a growing body of implementation guidance from federal agencies, standards bodies, and the vendor ecosystem. What it requires from managers is not technical mastery of quantum mechanics but operational clarity about what the organization holds, what protects it, and how to sequence the work of protecting it differently before the window closes. The chapters that follow provide the tools to do that work with precision and confidence.

Diagram 1.9 – The Post-Quantum Migration Imperative: A Managers Summary

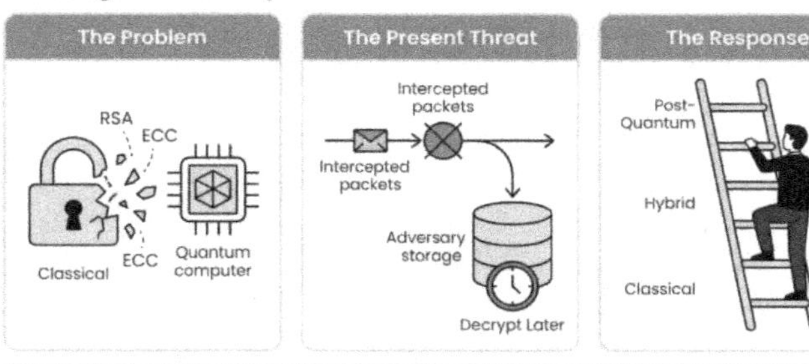

Diagram 1.9 – The Post-Quantum Migration Imperative: A Managers Summary

2　How the Machine Breaks the Lock: Quantum Algorithms and the Arithmetic of Risk

Scenario: A security architect at a large financial institution is preparing a briefing for the technology risk committee. She knows that quantum computing threatens RSA and ECC — she has read the headlines. What she does not know, and what the committee will certainly ask, is how specifically the threat works, what scale of quantum machine would be required, and which of the institution's cryptographic dependencies are actually at risk. She is technically literate but not a quantum physicist. She needs a working model of the threat that is precise enough to drive decisions without requiring a doctorate. This chapter is written for her and for every manager in a similar position.

Understanding how quantum algorithms attack classical cryptography does not require deep familiarity with quantum mechanics. It requires understanding two specific algorithms — Shor's and Grover's — at the level of what they do, what problem they solve, what they require to execute, and what the consequences are for specific cryptographic schemes. That understanding is the foundation for mapping the threat to your organization's specific cryptographic stack and for making defensible, mission-aligned decisions about which risks to address first and with what resources.

2.1 The Quantum Computational Model: What Makes It Different

Classical computers process information as bits — each bit is either zero or one. Every computation is a sequence of deterministic operations on those bits. Quantum computers process information as qubits, which can exist in superpositions of zero and one simultaneously. That property, combined with entanglement and interference, allows quantum computers to explore multiple computational paths at once and to extract structured information from that exploration in ways that classical computers cannot replicate. This is not magic and it is not indefinitely powerful. It is a specific set of physical properties that, when correctly harnessed, dramatically accelerates certain classes of computations while leaving others unchanged.

For managers, the critical insight is that quantum advantage is not universal. A quantum computer is not simply a faster classical computer. It is a different computational model that is dramatically faster for specific problem structures and no faster — or even slower — for others. The algorithms that underpin classical asymmetric cryptography happen to have exactly the structure that quantum computers can exploit. The algorithms that underpin symmetric cryptography have a different structure that is partially but not catastrophically affected. Understanding this distinction is what allows managers to make calibrated, proportionate responses rather than treating all of their cryptography as equally at risk.

Diagram 2.1 - Classical Bits vs. Quantum Qubits: Decision-Relevant Comparison

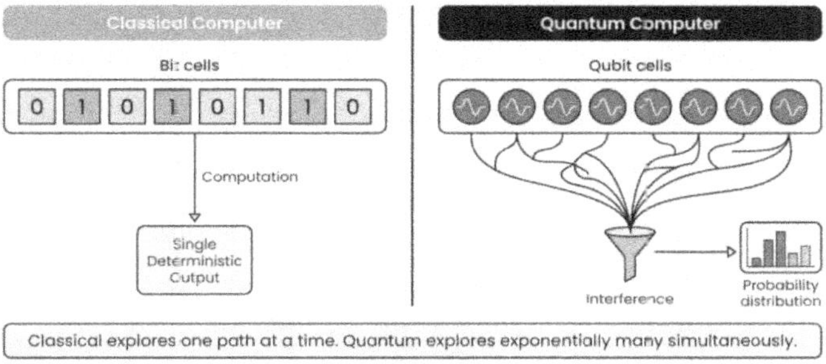

Diagram 2.1 - Classical Bits vs. Quantum Qubits: Decision-Relevant Comparison

2.1.1 Superposition, Entanglement, and Quantum Parallelism Without Magic

Superposition means that a qubit, unlike a classical bit, is not constrained to be either zero or one until it is measured. Before measurement, it exists in a probabilistic combination of both states. A quantum register of n qubits can represent 2^n states simultaneously. This does not mean the quantum computer stores all those states separately — that would require exponential memory. What it means is that the quantum computation operates on all states simultaneously through the mathematical structure of the quantum state. The key insight is that while a quantum computer can represent exponentially many states, it can only output one result when measured. The art of quantum algorithm design is configuring the computation so that interference patterns amplify the probability of measuring the correct answer while suppressing all other outcomes.

Entanglement is the property by which the quantum states of two or more qubits become correlated in ways that have no classical analog. When qubits are entangled, the state of one qubit is not independent of the others — measuring one qubit instantly determines the correlated properties of the others. Entanglement is the resource that allows quantum algorithms to create the interference patterns needed to amplify correct answers. It is a physical resource that must be engineered carefully in quantum hardware; it is also fragile, degrading through a process called decoherence as the system interacts with its environment. Managing decoherence is the central engineering challenge of building practical quantum computers.

2.1.2 Qubits Versus Classical Bits: A Decision-Relevant Comparison

For managers making decisions about post-quantum migration, the most useful framing of the qubit-versus-bit comparison is operational rather than physical. Classical bits are deterministic, stable, cheap to produce at scale, and extremely well understood after seventy years of development. Qubits are probabilistic, fragile, expensive to produce at any scale, and still the subject of active basic research. This operational contrast explains why quantum computers are not simply replacing classical computers: for the vast majority of computing tasks, classical hardware remains dramatically superior. Quantum hardware is superior only for the specific problem structures that quantum algorithms are designed to exploit.

The decision-relevant implication is that quantum computing's threat to cryptography is narrow but deep. The specific problems that quantum algorithms attack efficiently — factoring large integers, computing discrete logarithms in certain groups — happen to be the exact problems that the most widely deployed asymmetric cryptographic schemes depend on for their security. That overlap is not a consequence of cryptography's bad luck; it reflects that, before quantum computing was a practical concern, cryptographers selected mathematical problems that were hard for classical computers without necessarily being hard for all possible computational models. Post-quantum cryptography corrects that oversight by building on problem families that remain hard even under quantum computation.

2.1.3 Why Current Noise-Limited Machines Are Not Yet Cryptographically Relevant

Today's quantum computers operate in the Noisy Intermediate-Scale Quantum regime, commonly referred to as NISQ. NISQ devices have qubit counts ranging from tens to thousands of physical qubits, but their error rates are too high and their coherence times too short to execute the deep circuits required for cryptographic attacks on real-world key sizes. Shor's algorithm for RSA-2048, for example, requires executing millions of two-qubit gate operations on thousands of logical qubits. Logical qubits are error-corrected abstractions built from many physical qubits; achieving a single logical qubit requires hundreds to thousands of physical qubits depending on the error correction code and the physical error rate.

Current NISQ devices lack the scale and fidelity to implement error correction at the necessary depth. This is the gap between where the hardware is today and where it needs to be to constitute a cryptographically relevant quantum computer. The gap is real and significant. It is also closing through sustained engineering investment. The importance of the NISQ limitation for managers is that it provides time — not unlimited time, not a guarantee that the problem is distant, but a window of years in which migration can be planned and executed before the threat becomes operational. The critical mistake is treating that window as unlimited rather than bounded. A migration that takes seven years requires beginning seven years before the threat arrives.

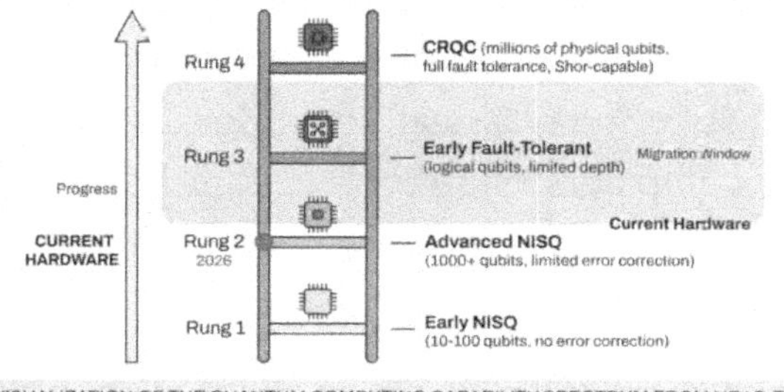

2.2 Shor's Algorithm: The Structural Threat to Public-Key Cryptography

Shor's algorithm, published in 1994, is the reason for the post-quantum migration. It provides a quantum algorithm that can factor large integers and compute discrete logarithms in polynomial time — a fundamentally different complexity class from the best-known classical algorithms, which require subexponential or exponential time for these problems at relevant key sizes. Polynomial time means the computation's runtime grows as a manageable power of the input size rather than exponentially with it. For RSA-2048, the difference between polynomial and exponential time corresponds roughly to the difference between solving the problem in hours and requiring more time than the universe has existed. Shor's algorithm collapses that difference.

The algorithm reduces the factoring problem to finding the period of a certain mathematical function — specifically, a function constructed from modular exponentiation. Finding that period classically is as hard as factoring itself. Quantum computers, via the Quantum Fourier Transform, can efficiently find this period. The algorithm's core insight is not brute force but structural: it finds a mathematical shortcut that classical computers cannot take because they cannot exploit quantum interference. The details of the quantum Fourier transform are not required knowledge for managers. Still, the strategic implication is essential: Shor's algorithm does not crack RSA by being fast at classical operations. It cracks RSA by exploiting a mathematical relationship that classical algorithms cannot reach.

2.2.1 Integer Factorization and Why RSA Trusted It

RSA was first published in 1977, and its security was grounded in a problem that had resisted efficient solution for centuries: given a large composite number $n = p \times q$ where p and q are large primes, find p and q. For large enough n — in modern practice, 2048 bits or more — the best classical algorithms require subexponential time, which is practically infeasible. The security of every RSA operation depends on this: the public key includes n and an exponent e, while the private key is derived from the factorization of n. If you know p and q, computing the private key is straightforward.

If you do not know them, and you cannot factor n, you cannot recover the private key.

RSA's designers were confident that integer factorization was hard because the mathematical community had worked on it intensively for centuries without finding an efficient classical algorithm. That confidence was and remains warranted for classical computation. The assumption they made — implicitly, since quantum computing did not yet exist as a concept — was that factoring would remain hard for any computational model. Shor's algorithm invalidated that assumption. It does not make classical factoring algorithms faster; it introduces a fundamentally different approach that bypasses the difficulty entirely. The lesson for enterprise architects is about the nature of cryptographic hardness assumptions: they are bounded by what was known about computation at the time, and that boundary has moved.

Diagram 2.3 – How Shors Algorithm Attacks RSA

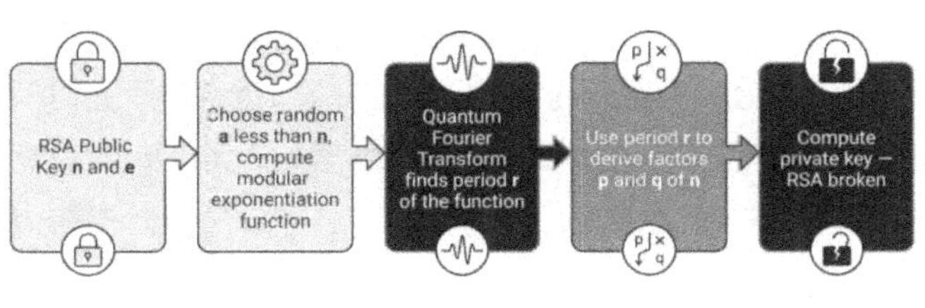

Caption – foreginn: classical nation is a step-by-step manager summary

2.2.2 Discrete Logarithms, Elliptic Curves, and Their Shared Vulnerability

Shor's algorithm does not only attack RSA. It provides an equally powerful attack against any cryptographic scheme whose security reduces to the difficulty of the discrete logarithm problem. Finite-field Diffie-Hellman key exchange, which underpins a significant portion of enterprise VPN configurations and older TLS deployments, relies on the discrete logarithm problem in multiplicative groups of integers modulo a prime. Elliptic Curve Cryptography relies on the elliptic-curve discrete logarithm problem — the same conceptual challenge, but defined over the algebraic structure of an elliptic curve, which allows shorter keys for equivalent classical security.

The shared vulnerability of these schemes to Shor's algorithm means that a single algorithmic advance has simultaneously exposed RSA, Diffie-Hellman, and ECC. These three schemes collectively underpin virtually all asymmetric cryptography currently deployed in enterprise environments: certificate authorities, TLS handshakes, SSH authentication, code signing, VPN tunnels, and identity federation. The scope of the exposure is not additive — it is comprehensive. Any system that uses asymmetric cryptography for key exchange or digital signatures and that uses one of these three schemes is in scope for migration. That is essentially every enterprise system that implements public-key cryptography.

2.2.3 What a Cryptographically Relevant
       Quantum Machine Would Actually Require

Executing Shor's algorithm on RSA-2048 requires an error-corrected quantum computer with millions of physical qubits organized into thousands of logical qubits capable of executing circuits with millions of T-gate operations. Current estimates of the physical qubit overhead required for logical qubit error correction, combined with circuit depth estimates for Shor on RSA-2048, produce a rough requirement of two to four million physical qubits operating with physical error rates significantly below current best-in-class hardware. This is a large gap from today's most capable systems, which operate in the hundreds to low thousands of physical qubits.

However, these estimates are sensitive to both algorithmic improvements and hardware advances. Researchers have repeatedly reduced the qubit-count estimates for Shor's algorithm through algorithmic optimization — new circuit compilation techniques, improved arithmetic circuits, and advances in error-correcting codes have each reduced the overhead. The hardware front shows similar progress: physical qubit counts are scaling, error rates are declining, and new qubit modalities are emerging with different error profiles. The convergence of algorithmic improvement and hardware advances is why the migration timeline is measured in years rather than decades. Managers should treat the qubit count estimates as illustrative of the gap rather than as a stable target.

2.3 Grover's Algorithm: The More Subtle Threat to Symmetric Encryption

Grover's algorithm, also published in the mid-1990s, provides a quadratic speedup for unstructured search problems. Where a classical computer searching an unsorted database of N items requires O(N) operations in the worst case, Grover's algorithm finds the target in O($\sqrt{N}$) quantum operations. This is a significant but not catastrophic speedup. For cryptographic applications, the relevant unstructured search is the brute-force key search: given a cryptographic cipher and some plaintext-ciphertext pairs, find the key. Grover's algorithm reduces the effective security margin of symmetric encryption schemes by a factor of $2^{(n/2)}$ in bits, because it can search $2^{(n/2)}$ candidates in the same time a classical computer would search 2^n.

This is the less dramatic but still important quantum threat to symmetric cryptography. It does not break symmetric encryption the way Shor's algorithm breaks RSA. The structure of symmetric ciphers — the way they mix key material into complex nonlinear transformations of data — does not provide the mathematical shortcut that Shor exploits. Grover's attack is a generic brute-force attack, just quadratically faster. The remediation is correspondingly straightforward: doubling the key length restores the pre-quantum security margin. The urgency is therefore calibrated differently for symmetric schemes than for asymmetric ones.

Diagram 2.4 – Grovers Quadratic Speedup on Symmetric Key Search

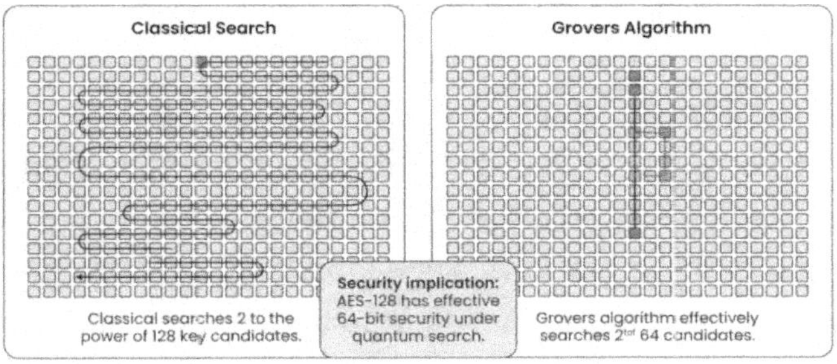

Diagram 2.4 – Grovers Quadratic miniml 64-bit security under quantum search

2.3.1 Quadratic Speedup and What It Means for Key-Length Security

The quadratic speedup means that a symmetric key of n bits, which classically requires 2^n operations to brute-force, requires only approximately $2^{(n/2)}$ quantum operations under Grover's algorithm. This halves the effective key length in bits. A 128-bit symmetric key, which classically provides 128 bits of security, provides only approximately 64 bits of effective security against a quantum adversary using Grover's algorithm. Sixty-four bits of security is insufficient: it is within reach of a well-resourced classical adversary using distributed computing, let alone a future quantum adversary. A 256-bit key, by contrast, provides approximately 128 bits of effective post-quantum security, which is the current consensus minimum for post-quantum cryptographic schemes and is considered adequate for the foreseeable future.

The practical implication for enterprise managers is that AES-256 is the symmetric encryption standard to standardize on, and AES-128 should be phased out in systems where symmetric security is a direct requirement. This is a relatively tractable migration compared to the asymmetric migration: AES-256 is already supported by virtually all modern cryptographic libraries and hardware acceleration units. The constraint is not technical availability but configuration discipline — ensuring that cipher suite selections across TLS configurations, database encryption settings, full-disk encryption deployments, and backup encryption policies consistently use 256-bit keys. Configuration audits and policy enforcement tooling can address this systematically.

2.3.2 AES-128 Versus AES-256 in a Post-Quantum World

AES-128 and AES-256 are both variants of the Advanced Encryption Standard, which is a block cipher that has withstood extensive classical cryptanalysis since its standardization in 2001. The key lengths differ — 128 bits versus 256 bits — but both variants use the same core algorithm with different numbers of processing rounds (10 rounds for AES-128, 14 for AES-256). In the classical threat model, AES-128 provides security that is indistinguishable from AES-256 for practical purposes: no classical attack comes close to the 2^{128} operations required for a brute-force key search. In the post-quantum model, Grover's algorithm changes the calculus specifically for brute-force key search — and only for that attack. It does not affect the structural security of the AES algorithm itself.

The verdict for enterprise deployment is clear: standardize on AES-256 for all new deployments and migrate existing AES-128 deployments during the post-quantum transition period. This is not an emergency remediation — AES-128 does not become immediately broken by Grover's algorithm. It becomes inadequate at the security margin level for long-horizon data protection. For a manager conducting a phased migration, symmetric key length remediation is a relatively low-friction task that can be incorporated into existing patching and configuration management cycles, unlike the asymmetric migration, which requires coordinated protocol changes across multiple system boundaries.

2.3.3 Hash Functions and Collision Resistance Under Quantum Search

Hash functions — which produce fixed-length digests from arbitrary input data — serve multiple cryptographic purposes: integrity verification, digital signature construction, key derivation, and password hashing. Their security properties include preimage resistance (given a hash output, it is hard to find any input that produces it), second preimage resistance (given an input and its hash, it is hard to find a different input with the same hash), and collision resistance (it is hard to find any two distinct inputs with the same hash). Grover's algorithm improves preimage resistance by providing a quadratic speedup in preimage finding. Quantum collision-finding algorithms provide a smaller speedup in collision resistance — roughly a cubic-root acceleration rather than a square-root — under certain conditions.

For enterprise applications, the practical guidance is to use SHA-256 or SHA-3-256 at a minimum and prefer SHA-384 or SHA-512 for applications requiring long-horizon security. SHA-1 and MD5 are already deprecated on classical grounds and should not appear in any post-quantum migration planning as acceptable options — they must be treated as immediate remediation items regardless of quantum considerations. The hash function migration is typically less disruptive than the asymmetric migration and should be included in the configuration baseline that the migration program establishes in its early phases.

2.4 Mapping Algorithmic Threats to Your Cryptographic Stack

The academic treatment of Shor's and Grover's algorithms is a useful background. Still, the decision a manager needs to make is concrete: which of my organization's deployed protocols are broken, weakened, or unaffected? That mapping is what converts theoretical risk into a prioritized action plan. The answer is structured around three categories: schemes that Shor's algorithm breaks entirely (asymmetric schemes based on factoring and discrete logarithms), schemes that Grover's algorithm weakens but does not break (symmetric schemes with insufficient key lengths), and schemes that neither algorithm significantly affects (correctly parameterized symmetric encryption and hash functions).

Diagram 2.5 – Quantum Algorithm Threat Map

BROKEN — Shor-Capable	WEAKENED — Grover-Affected	UNAFFECTED — Quantum Safe
RSA	AES-128 (64-bit effective)	AES-256
Diffie-Hellman	SHA-256 (reduced preimage)	SHA-384
ECDH		SHA-512
ECDSA		HMAC-256
DSA		
Migrate Now	**Upgrade to 256-bit**	**No Immediate Action**

Caption al clean minimlist infographic

2.4.1 Which Deployed Protocols Are Broken, Weakened, or Unaffected

The broken category is definitive and large. Every protocol that uses RSA for key encapsulation or digital signatures is broken by a CRQC running Shor's algorithm. This includes TLS cipher suites that use RSA key exchange (RSA-PKCS1 and RSA-OAEP-based suites), SSH sessions authenticated with RSA host keys or user keys, S/MIME and PGP messages signed or encrypted with RSA keys, code signing certificates issued with RSA keys, and X.509 certificates in your PKI hierarchy that use RSA signature algorithms. Every protocol that uses elliptic curve cryptography for key agreement or signatures — ECDH, ECDSA, EdDSA — is equally broken by Shor's algorithm, because the elliptic curve discrete logarithm problem is as susceptible as the integer factorization problem.

The weakened category includes symmetric encryption with 128-bit keys and hash functions with 160-256-bit outputs, assuming that Grover's algorithm yields a practical speedup on realistic hardware scales. The practical significance depends on the data's confidentiality horizon and the resources available to a future quantum adversary. For most enterprise applications, AES-128-encrypted data today is not at immediate risk from Grover-enhanced attacks — the quadratic speedup is real, but the absolute computation required remains substantial. The priority distinction is between weakened (requires eventual remediation within the migration program) and broken (requires urgent remediation before asymmetric attacks become feasible).

2.4.2 TLS, PKI, Code Signing, and VPNs: An Algorithm-by-Algorithm Assessment

TLS 1.3, the current standard for transport layer security, uses forward secrecy by default through ephemeral key exchange. The key exchange is typically performed using X25519, which uses the Curve25519 elliptic-curve Diffie-Hellman, or with finite-field groups. Shor's algorithm breaks both. However, TLS 1.3's forward secrecy property means that a recorded TLS session cannot be decrypted by compromising the server's long-term private key — the session key was ephemeral. This is a meaningful distinction: it means that TLS 1.3 sessions recorded today would require the quantum attacker to run Shor's algorithm on the ephemeral key material from each session individually, rather than compromising a single long-term key and decrypting all historical sessions. This property was intentional and provides partial protection, but it does not eliminate the harvest now threat — it only raises the per-session cost of decryption.

PKI infrastructure is the deepest layer of vulnerability. Certificate authorities issue certificates by signing them with their private key — currently RSA or ECDSA in virtually all deployments. The root CA's signing key is the root of trust for the entire enterprise certificate infrastructure. A quantum attacker who can compute the private key from the public key in a root CA certificate can forge certificates for any domain or service in the enterprise and impersonate any authenticated entity. PKI migration must therefore begin at the root and proceed downward: root CAs migrated first, intermediate CAs second, end-entity certificates last. This sequencing constraint means that PKI migration is one of the longest-lead items in the enterprise migration program and must be started early.

Code signing poses a specific risk worth calling out explicitly. Signed software — operating system components, firmware updates, application installers — carries a cryptographic guarantee of authenticity. If a quantum attacker can forge the signing key used to sign current production software, they can create malware that appears cryptographically authentic. The same principle applies to VPN configurations: VPN gateways authenticate to clients using certificates, and those certificates carry signatures that a quantum attacker could potentially forge. This is not a near-term risk given current hardware limitations. Still, it is a use case where the confidentiality horizon of the key material and the consequences of forgery make early migration particularly important.

2.4.3 Symmetric Versus Asymmetric Remediation Priorities

The remediation priority distinction between symmetric and asymmetric cryptography is one of the most operationally important points in this chapter for managers setting migration scope and schedule. Asymmetric schemes — RSA, DH, ECC — are entirely broken by Shor's algorithm and must be replaced with post-quantum alternatives. No parameterization of RSA survives quantum attack; using a longer RSA key buys more time against classical attackers but not against quantum attackers, because Shor's polynomial-time algorithm scales to any RSA key size that is computationally accessible. The remediation is migration to a fundamentally different algorithm family, not a parameter change.

Symmetric schemes require remediation only in their key length parameterization. Doubling the key length from 128 to 256 bits restores the security margin to pre-quantum levels. This is a configuration change, not an algorithm replacement, and it is much less disruptive. The migration priority ranking for most enterprises is therefore: first, asymmetric schemes in high-criticality applications (PKI, authentication, key exchange protocols); second, asymmetric schemes in all remaining applications; third, symmetric key length upgrades in applications where AES-128 is deployed with long-horizon data. This sequencing reflects actual risk, not operational convenience, and it is the basis for defending resource allocation to leadership.

Diagram 2.6 – Asymmetric vs. Symmetric Remediation Priority Matrix

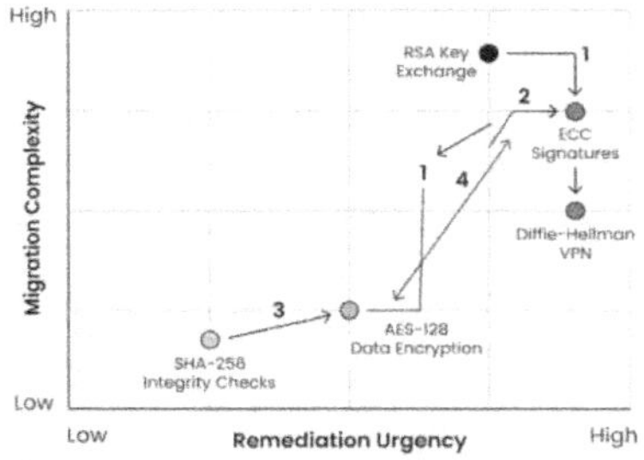

Diagram 2.6 – Asymmetric vs. Symmetric Remediation Priority Matrix

2.5 Beyond Shor and Grover: Emerging Quantum Attack Vectors

Shor's and Grover's algorithms represent the known, well-characterized quantum threats to classical cryptography. They are the basis for NIST's post-quantum standardization process, the federal migration mandates, and the enterprise planning frameworks now being developed across multiple sectors. But quantum computing is an active research field, and the threat landscape is not static. Managers who build migration programs exclusively around Shor and Grover, without establishing a mechanism for monitoring emerging developments, are building a program calibrated to today's known threats rather than tomorrow's evolving ones. This section surveys the additional considerations that a responsible migration program should take into account.

2.5.1 Quantum-Assisted Cryptanalysis of Non-Standard Constructions

Post-quantum cryptographic standards, which are the replacement for RSA and ECC, are designed to resist both classical and quantum attacks on their core mathematical problems. But enterprises do not run only standard algorithms. They run custom protocol implementations, proprietary cryptographic constructions, legacy systems with non-standard cipher suite configurations, and hybrid schemes that combine standard algorithms in non-standard ways. Each of these represents a surface area that may not have been specifically analyzed for quantum resistance.

Quantum computing may assist in analyzing these non-standard constructions in ways that are not yet fully characterized. Quantum algorithms for solving linear systems (the HHL algorithm), quantum random walk algorithms, and quantum-enhanced machine learning approaches have been proposed as potential tools for attacking cryptographic constructions that do not fit the standard factoring or discrete logarithm structures. The current consensus is that these approaches do not threaten well-designed standard algorithms, but they may accelerate cryptanalysis of weak or non-standard constructions. The migration program implies that non-standard cryptographic implementations should be identified and replaced with standards-compliant alternatives as part of the migration, not preserved because they do not obviously match the Shor/Grover threat profile.

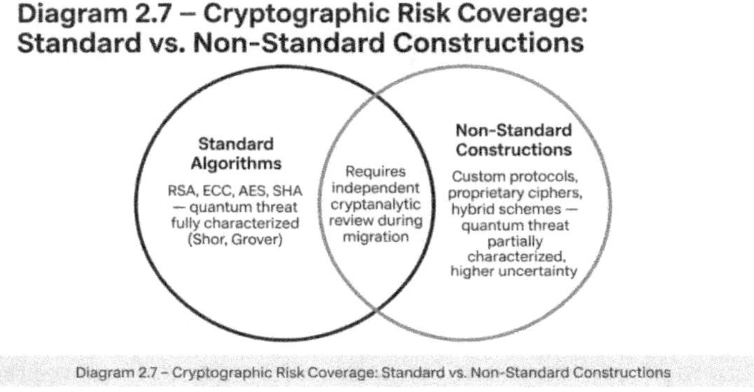

Diagram 2.7 – Cryptographic Risk Coverage: Standard vs. Non-Standard Constructions

2.5.2 Side-Channel and Fault-Injection Risks in Quantum-Era Hardware

The post-quantum algorithms that NIST has standardized are mathematically resistant to quantum attacks on their core problem structure. But mathematical resistance to quantum attacks does not automatically guarantee resistance to side-channel and fault-injection attacks on the physical implementations of those algorithms. Side-channel attacks extract secret key material by observing physical characteristics of the computation — timing variations, power consumption patterns, electromagnetic emissions —

rather than attacking the mathematical structure directly. Fault-injection attacks induce computational errors in cryptographic hardware to recover key material or bypass authentication checks.

Post-quantum algorithms present new side-channel challenges compared to their classical counterparts. Lattice-based algorithms, for example, involve operations on polynomial coefficients that have different power and timing profiles than RSA or ECC operations. Some early implementations of post-quantum algorithms are vulnerable to timing attacks that were not present in their classical counterparts. Hardware security modules, smart cards, and embedded cryptographic processors must therefore undergo post-quantum algorithm validation not only for mathematical correctness but for side-channel resistance in their specific hardware context. Enterprise procurement processes should require that vendors demonstrate side-channel resistance testing for post-quantum algorithm implementations, not just algorithmic compliance. This is a workflow-level impact consideration that belongs in vendor evaluation frameworks from the earliest phases of the migration.

2.6 Manager's Checklist: Chapter 2

**Diagram 2.8 – Algorithm Threat Classification:
A Managers Reference**

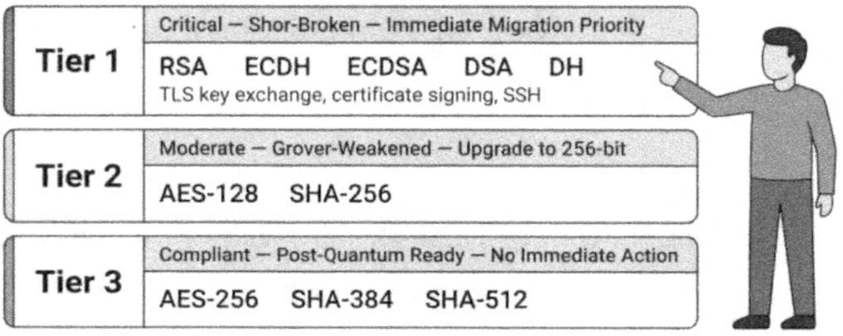

Tier 1	Critical – Shor-Broken – Immediate Migration Priority
	RSA ECDH ECDSA DSA DH
	TLS key exchange, certificate signing, SSH

Tier 2	Moderate – Grover-Weakened – Upgrade to 256-bit
	AES-128 SHA-256

Tier 3	Compliant – Post-Quantum Ready – No Immediate Action
	AES-256 SHA-384 SHA-512

Diagram 2.8 – Algorithm Threat Classification: A Managers Reference

- Confirm whether your security team has mapped which deployed protocols rely on RSA, DH, or ECC for key exchange or signatures. If this mapping does not exist, initiate a cryptographic discovery exercise.
- Identify any applications or infrastructure components in your environment that use AES-128 as the primary symmetric encryption scheme for data with confidentiality horizons exceeding five years.
- Verify that your TLS 1.3 deployment uses forward secrecy (ephemeral key exchange) by default and that TLS 1.2 and earlier are either disabled or restricted to legacy compatibility contexts only.
- Request vendor documentation confirming whether each of your major infrastructure vendors — HSMs, network equipment, identity platforms, cloud providers — has published post-quantum algorithm roadmaps and implementation timelines.
- Confirm that code signing certificates and the pipelines that use them are included in the migration

scope. Code signing is often omitted from initial migration planning and is a high-value target.

- Determine whether your organization has any custom or proprietary cryptographic implementations that are not based on NIST-standardized algorithms. These require an independent security review as part of the migration, not just algorithm substitution.

- Ensure that your migration planning accounts for side-channel resistance requirements in HSM and embedded cryptographic hardware — mathematical compliance with post-quantum algorithms is necessary but not sufficient for hardware security components.

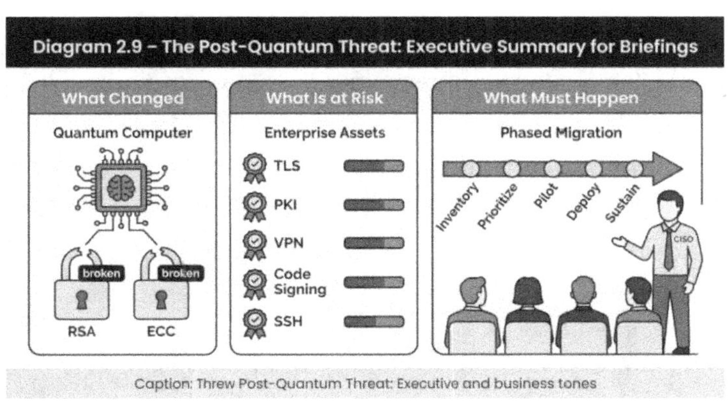

Caption: Threw Post-Quantum Threat: Executive and business tones

2.7 The Arithmetic Is Not Your Enemy — Ignorance of It Is

Shor's algorithm is the reason asymmetric cryptography has an expiration date. Grover's algorithm is the reason symmetric key lengths need to be revisited. Together, they define a threat surface that is larger and more urgent than most enterprises have formally scoped. But the threat is characterized. It is not a mystery wrapped in speculation — it is a specific pair of algorithms with known computational requirements, known target algorithms, and known remediation paths.

What this chapter provides is the basis for making risk-proportionate decisions. Not every piece of cryptography in your enterprise is equally at risk, and not every remediation is equally urgent. The asymmetric migration is the priority. The remediation of symmetric and hash functions is real, but more tractable and less time-constrained. The emerging threat landscape — quantum-assisted cryptanalysis of non-standard constructions, side-channel attacks on post-quantum implementations — is the horizon that a well-designed migration program builds monitoring capacity to watch. Mission-aligned security requires understanding not just that the threat exists but exactly where it bites and how hard, so that the response is calibrated to the actual exposure rather than to generalized anxiety about a technology most managers have never seen operate.

The next chapter turns from the nature of the threat to the nature of the solution: the post-quantum algorithm families that NIST has standardized, what makes them resistant to quantum attacks, and what their deployment characteristics mean for enterprise architecture decisions.

3 Hardness That Survives the Quantum Age: The New Algorithmic Landscape

Scenario: The security architecture team at a major healthcare network has just completed its first cryptographic asset inventory. The results are sobering: RSA-2048 certificates throughout the PKI, ECDH key exchange in TLS configurations, and a mix of SHA-256 and deprecated SHA-1 in legacy application integrations. The CISO has been told to transition to post-quantum cryptography, and she has read the relevant NIST guidance. But now the team faces the question that the guidance does not fully answer: which post-quantum algorithm family should we choose, and for which of our use cases? The CISO needs to understand not just that post-quantum algorithms exist, but also why they work, how they compare, and what their deployment characteristics mean for a healthcare network with constrained change-management bandwidth and tight availability requirements.

This chapter provides that understanding. It surveys the four principal families of post-quantum cryptographic constructions — lattice-based, hash-based, code-based, and multivariate polynomial — along with the cautionary lessons of isogeny-based cryptography. For each family, it explains the underlying hard problem at an intuitive level, the security argument for quantum resistance, the performance and key-size characteristics that determine deployment fit, and the use cases where each family is most appropriate. The chapter closes with a framework for matching algorithm families to specific enterprise use cases based on performance requirements, key size constraints, operational overhead, and risk tolerance.

3.1 What Makes a Problem Quantum-Resistant: The Right Kind of Hard

The post-quantum migration is not simply a matter of swapping one set of algorithms for another. It is a matter of replacing algorithms that quantum computers can solve efficiently with algorithms for problems they cannot. Understanding which problems have that property — and why — is the foundation for evaluating the algorithm families and the standards built on them. It is also the basis for applying appropriate skepticism when new algorithm proposals appear and for understanding why the cryptographic community's confidence in specific post-quantum constructions varies significantly by family and by the length of time the underlying problem has been studied.

3.1.1 Why Quantum Advantage Does Not Apply Uniformly Across Problem Classes

Shor's algorithm works by finding the period of a specific mathematical function related to the target problem — factoring or discrete logarithm. This period-finding approach is powerful because the problems it solves have a hidden algebraic structure — an abelian group structure — that the quantum Fourier transform can exploit efficiently. Not all hard computational problems have this structure. Problems that lack hidden algebraic periodicity are not susceptible to Shor-type algorithms. The challenge for post-quantum cryptography is to identify problems that are demonstrably hard for both classical and quantum adversaries and to build cryptographic constructions on those problems that are also efficient enough for practical deployment.

Grover's algorithm is more general — it provides a quadratic speedup for any unstructured search problem — but a quadratic speedup is manageable by doubling key sizes. Grover-type attacks do not primarily threaten the families of problems that post-quantum cryptography builds on; they are threatened by potential classical cryptanalytic advances and by any quantum algorithm specifically designed for their problem structure. The current consensus, after years of public cryptanalytic scrutiny through the NIST standardization process, is that lattice-based, hash-based, code-based, and multivariate problems do not have the algebraic structure that Shor exploits, and do not succumb to any known quantum algorithm more efficiently than classical ones beyond Grover's generic speedup. That consensus is the basis for confidence in the post-quantum standards — not certainty, but calibrated confidence built on sustained public analysis.

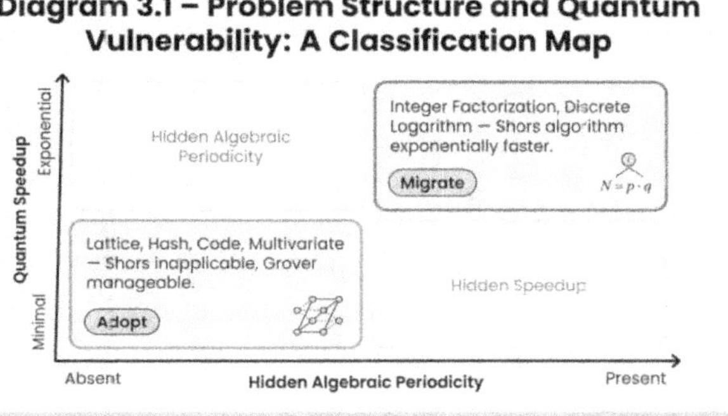

Diagram 3.1 – Problem Structure and Quantum Vulnerability: A Classification Map

3.1.2 The Role of Worst-Case to Average-Case Reductions in Security Proofs

One of the most technically important properties of lattice-based cryptography — and the reason it has attracted so much attention from the cryptographic research community — is the existence of worst-case to average-case security reductions for certain lattice problems. In classical cryptography, most security arguments are "average-case": they assert that a randomly selected instance of the problem is hard, relying on empirical evidence rather than mathematical proof. A worst-case to average-case reduction provides a stronger guarantee: it proves that if the cryptographic scheme can be broken, then the underlying hard problem can also be solved in its worst case — the absolute hardest instances of the problem. This is a significantly stronger security argument.

For managers, the practical implication of this property is about confidence calibration. Lattice-based schemes with strong worst-case reductions provide security arguments that are more formally grounded than those of competing families. If the cryptographic scheme is broken in practice, the mathematical proof guarantees that an efficient algorithm for the underlying lattice problem also exists, which would represent a fundamental advance in computational complexity theory rather than just a cryptanalytic finding. This does not make lattice-based schemes unbreakable, but it means that breaking them in an unexpected way is substantially harder than finding a practical attack on a scheme without such reductions. For procurement decisions and risk assessments, this property is worth understanding even if its technical details remain opaque.

3.1.3 Evaluating Newness: How Long Has the Problem Been Studied?

The history of cryptography is littered with algorithms that appeared secure until they were not. A recurring theme in cryptographic failures is insufficient public scrutiny before deployment: when an algorithm is deployed before the research community has had time to probe its assumptions thoroughly, subtle weaknesses may go undetected until an adversary exploits them or a researcher publishes a break. The NIST post-quantum standardization process was explicitly designed to address this by running an open, multi-year competition that invited the global research community to submit, analyze, and attack candidate algorithms. That process ran for approximately six years and produced the current standards — ML-KEM, ML-DSA, and SLH-DSA.

For a manager evaluating algorithm families, the age and depth of the underlying hard problem's study history is a meaningful risk indicator. Lattice problems — specifically the Learning With Errors and Short Integer Solution problems — have been studied since the mid-1990s and have attracted significant cryptanalytic attention with no polynomial-time classical or quantum algorithm found. Hash functions have been studied for decades with an extensive body of analysis. Code-based cryptography, which builds on error-correcting codes, has been studied since 1978 — longer than RSA. Multivariate polynomial problems have a shorter track record of cryptographic deployment. Isogeny-based cryptography was the newest and least-studied family, and its most prominent scheme was broken in 2022, even though standardization appeared imminent. The lesson is that novelty in a cryptographic construction is a risk factor, not a selling point.

3.2 Lattice-Based Cryptography: Geometry as a Security Foundation

Lattice-based cryptography is the foundation of the primary NIST post-quantum standards: ML-KEM (formerly CRYSTALS-Kyber) for key encapsulation and ML-DSA (formerly CRYSTALS-Dilithium) for digital signatures. Understanding why lattices provide a sound foundation for cryptography — and what their deployment characteristics mean for enterprise architecture — is essential for any manager responsible for evaluating or implementing post-quantum migration plans. The underlying mathematics is more complex than RSA, but the security intuition is accessible, and the performance characteristics are concrete.

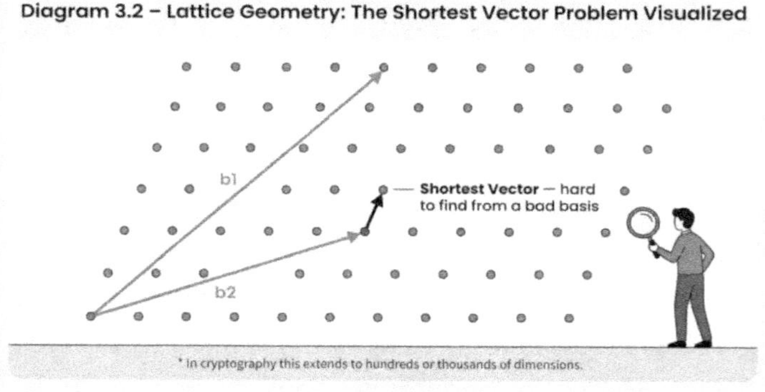

Diagram 3.2 – Lattice Geometry: The Shortest Vector Problem Visualized

3.2.1 The Shortest Vector and Learning With Errors Problems

A lattice is a set of evenly spaced points in high-dimensional space, generated by linear combinations of a set of basis vectors with integer coefficients. The geometric intuition is a grid: in two dimensions, think of a checkerboard pattern. In cryptographically relevant dimensions — hundreds or thousands of dimensions — the geometric structure is far more complex and computationally opaque. The Shortest Vector Problem asks: given a basis for a lattice, find the lattice vector with the smallest Euclidean length. For a carefully chosen lattice in high dimensions, this problem is computationally hard for classical computers and, crucially, no efficient quantum algorithm for it is known.

The Learning With Errors (LWE) problem is a specific lattice problem that has become the foundation for the most widely deployed post-quantum cryptographic schemes. LWE asks: given a collection of approximate linear equations over integers modulo a prime, with each equation perturbed by a small random error, recover the secret vector that satisfies them exactly. The hardness of LWE in high dimensions is provably related to the hardness of the Shortest Vector Problem through worst-case to average-case reductions. This formal reduction is the source of the strong security guarantees that lattice-based schemes can provide.

ML-KEM and ML-DSA are both built on variants of LWE and the related Short Integer Solution problem.

3.2.2 Why Lattices Are Believed to Resist Quantum Attacks

The resistance of lattice problems to quantum attacks rests on the absence of hidden algebraic structure that Shor-type algorithms can exploit. Lattices do not have the abelian group structure that the quantum Fourier transform can interrogate efficiently. Quantum algorithms for lattice problems have been extensively studied, and the best-known approaches provide no significant asymptotic speedup over classical algorithms on the problem instances used in cryptographic constructions. Grover's generic search speedup applies to lattice-based schemes in the same way it applies to all symmetric-equivalent problems: it reduces the security by half the key-equivalent bits, which is addressed by choosing parameters that provide sufficient security margin even after accounting for this reduction.

The academic research community has worked for three decades to find efficient quantum algorithms for lattice problems, motivated by both the security implications and the fundamental questions of computational complexity involved. No polynomial-time or subexponential quantum algorithm has been found for the Shortest Vector Problem or LWE at the dimensions used in cryptographic constructions.

This does not prove that none exists — mathematical proofs of hardness remain an open challenge in computational complexity theory — but it represents a body of negative evidence that is substantially deeper and more recent than the evidence supporting the RSA and ECC assumptions. Quantum resistance is therefore a confident engineering judgment, not a mathematical theorem, but it is the most strongly supported confident judgment available.

3.2.3 Performance Characteristics and Key-Size Implications for Enterprise Systems

Lattice-based schemes, particularly ML-KEM, offer performance characteristics competitive with classical asymmetric schemes in most enterprise deployment contexts. Key and ciphertext sizes are larger than those for ECC — ML-KEM-768, the recommended parameter set for most applications, which uses public keys of approximately 1184 bytes and ciphertexts of approximately 1088 bytes, compared to 64 bytes for an X25519 public key. This is a roughly 18-fold increase in key material size. For most enterprise applications — certificate issuance, TLS handshakes, SSH key exchange — this size increase is accommodated without significant operational impact. Network overhead per handshake increases by a few kilobytes, which is negligible on modern enterprise networks. Memory requirements for software implementations are well within the range of standard server hardware.

The computational performance of lattice-based schemes is generally favorable. Operations per second for key generation, encapsulation, and decapsulation in ML-KEM are comparable to, or faster than, those of RSA-2048 in well-optimized software implementations. Digital signature operations for ML-DSA are somewhat slower than ECDSA for signing, due to the rejection-sampling approach used in the signature algorithm, but verification is fast. For high-throughput use cases — TLS termination at a large-scale load balancer, high-volume API signing — performance benchmarking should be conducted in the target deployment environment to confirm that the throughput requirements are met. For the majority of enterprise use cases, lattice-based performance is adequate, and adoption is the recommended default.

3.2.4 Where Lattice Schemes Excel and Where They Introduce Friction

Lattice-based schemes excel in the core enterprise use cases that require both key encapsulation and digital signatures: TLS handshakes, certificate-based authentication, email encryption, and code signing. The combination of strong security foundations, reasonable performance, and availability in the NIST standards makes them the default choice for enterprises beginning post-quantum migration. They are also the algorithm family with the most mature and extensively reviewed implementation ecosystem, with open-source libraries available for all major programming languages and hardware acceleration support emerging across multiple platform vendors.

The friction points for lattice-based schemes appear in resource-constrained environments. Embedded systems with limited flash memory and RAM — IoT sensors, industrial controllers, smart card chips, and certain HSM firmware configurations — may struggle to accommodate the larger key and ciphertext sizes. The rejection sampling in ML-DSA signing introduces non-deterministic timing behavior, which requires careful handling in constant-time implementations to avoid timing side channels. For constrained devices, the trade-off analysis between lattice-based schemes and more compact alternatives — particularly hash-based signatures — is worth conducting explicitly rather than defaulting to a one-size-fits-all choice. The migration program should segment its asset population by resource constraints and apply different algorithm selections accordingly.

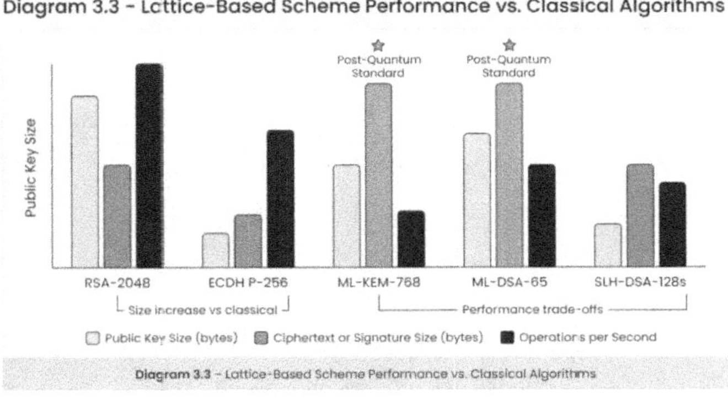

Diagram 3.3 – Lattice-Based Scheme Performance vs. Classical Algorithms

3.3 Hash-Based Signatures: Conservatism as a Feature

Hash-based digital signature schemes occupy a unique position in the post-quantum landscape: their security depends only on the collision resistance and preimage resistance of the underlying hash function. If SHA-256 or SHA-3 is secure, then properly parameterized hash-based signatures are secure. This simplicity is a feature, not a limitation. It means that hash-based schemes carry the deepest, most straightforward security argument of any post-quantum family: the same confidence you have in SHA-3 is directly inherited by the signature scheme built on it. For risk-averse organizations and high-value use cases, this conservative foundation is genuinely valuable.

3.3.1 One-Time and Few-Time Signature Schemes and Their Statefulness Constraints

The foundational hash-based signature schemes are one-time: a given key pair can safely sign exactly one message. Signing a second message with the same key reveals information that could allow an attacker to forge signatures. This statefulness constraint is not a hypothetical risk — it is a mathematically precise limitation. Early hash-based schemes addressed this by constructing trees of one-time key pairs (Merkle trees), in which the root of the tree serves as the long-term public key, and individual leaf keys are used for individual signatures. The state that must be maintained is a counter tracking which leaf keys have been used, ensuring that no leaf key signs more than one message.

The statefulness requirement is the primary operational challenge of hash-based signatures. In enterprise environments, statefulness means that signing operations must coordinate to prevent key reuse — two processes cannot sign with the same leaf key without risking security. In distributed systems, high-availability deployments, or systems that recover from crashes, maintaining signature state reliably is an engineering challenge that requires careful implementation. Statefulness is manageable but must be explicitly designed for. Organizations that deploy hash-based signatures without implementing robust state management will create security vulnerabilities that may not be immediately visible. This is the primary reason that hash-based schemes are recommended for specific use cases — particularly those with low signing rates and centralized signing infrastructure — rather than as a universal replacement for all signature applications.

3.3.2 Stateless Hash-Based Signatures and the SPHINCS+ Approach

SPHINCS+, standardized as SLH-DSA (Stateless Hash-Based Digital Signature Standard), addresses the statefulness problem through a clever construction that eliminates the need to track signing state while preserving the hash-function-only security foundation. The scheme uses a hypertree structure — a tree of trees — combined with a few-time signature scheme in the lower layers. By generating the signing key for each message by hashing both the secret key and a randomized message digest, SPHINCS+ avoids any deterministic relationship between messages and leaf positions that would require state tracking. Each signature is effectively self-contained: the verifier can reconstruct the verification path without reference to any external state.

The operational consequence of statelessness is significant. SLH-DSA can be deployed in distributed systems, high-availability configurations, and crash-recovery scenarios without the engineering overhead of synchronized state management. The trade-off is signature size and signing speed. SLH-DSA-128s — the smallest and slowest variant in the standard — produces signatures of approximately 7,856 bytes and is meaningfully slower than ML-DSA in signing operations. SLH-DSA-128f — the fast variant — uses larger signatures (17,088 bytes) but achieves faster signing. These sizes are substantially larger than both classical ECDSA signatures and ML-DSA signatures. For bandwidth-sensitive applications or high-throughput signing pipelines, this is a meaningful operational constraint. For low-frequency, high-value signing operations — root CA

certificate issuance, firmware signing, document signing —
SLH-DSA is an excellent choice precisely because of its
conservative security foundation and stateless operation.

Diagram 3.4 – SPHINCS+ SLH-DSA Signature Tree Structure

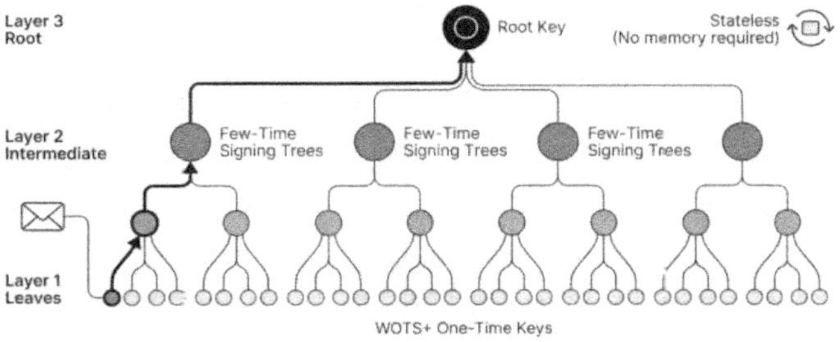

Caption: – SPHINCS+ SLH-DSA Signature Tree Infographic. Simplified representation of the nested trees within the SPHINCS+ scheme.

3.3.3 Use Cases Where Hash-Based Schemes Are the Right Default

Hash-based signatures are the right default for use cases
characterized by low signing frequency, high security
requirements, conservative trust environments, and
tolerance for large signature sizes. Root and intermediate
certificate authority signing operations fit this profile
exactly: a root CA signs a small number of certificates per
year; each certificate-signing operation is a high-value
event; the environment can afford the computation time; and
the signature is embedded in a certificate that is not
bandwidth-constrained. SLH-DSA is an appropriate choice

for root CA signing keys and is being evaluated for this use case by multiple certificate authority operators.

Firmware signing is another strong fit. Firmware updates are infrequent, signing is centralized, signature size is not a significant operational constraint relative to the firmware payload, and the security consequences of a forged firmware signature are severe — a successful forgery could compromise millions of deployed devices. The hash-function-only security argument is particularly compelling for firmware signing because the simplicity of the security foundation reduces the attack surface and the dependency on assumptions about problems that future mathematical advances might break. Code signing certificates in software supply chain pipelines, time-stamping authorities, and audit log integrity systems are additional use cases in which hash-based signatures combine appropriate performance characteristics with the strongest available security guarantees.

3.4 Code-Based and Multivariate Approaches: Diversity in the Post-Quantum Portfolio

Lattice-based and hash-based schemes are the primary post-quantum standards for most enterprise use cases. But a responsible post-quantum portfolio does not rely on a single algorithm family. Algorithm diversity is a deliberate risk-management strategy: if a fundamental mathematical advance or cryptanalytic discovery undermines one family, systems using other families remain secure. Code-based and multivariate polynomial approaches represent alternative families with distinct security assumptions, performance characteristics, and deployment strengths. Understanding them is important not because enterprises should deploy them universally, but because a defense-in-depth posture includes knowing when and where to apply them.

3.4.1 Error-Correcting Codes as a Cryptographic Primitive

Code-based cryptography builds on the theory of error-correcting codes — specifically, on the difficulty of decoding a general linear code. Error-correcting codes are mathematical structures used to detect and correct errors introduced during data transmission or storage. Given a received codeword that has been corrupted by errors, decoding the original message is computationally hard for a general random code, even though efficient decoding algorithms exist for specific structured codes. The McEliece cryptosystem, first proposed in 1978, exploits this asymmetry: the public key is a disguised form of a structured code whose efficient decoding algorithm is hidden, while the private key contains the decoding algorithm. Encrypting a message means deliberately introducing errors into the encoded message; decrypting requires the decoding algorithm.

The McEliece system has the longest security track record among post-quantum schemes. It has been studied continuously since 1978 with no polynomial-time classical or quantum attack found against the core problem — decoding a random linear code. This longevity is genuinely impressive and distinguishes it from newer constructions. The practical limitation that has prevented widespread deployment is key size: the public key in the original McEliece construction is very large — megabytes for cryptographic security levels — because the code's disguised matrix is transmitted as the public key. NIST evaluated Classic McEliece as a fourth-round candidate and acknowledged its security track record. Still, its key size makes it impractical for many enterprise use cases without specific engineering accommodations. For environments where large key sizes are acceptable — batch key distribution systems, long-term archival encryption, certain classified data protection contexts — code-based schemes offer a security argument of exceptional longevity.

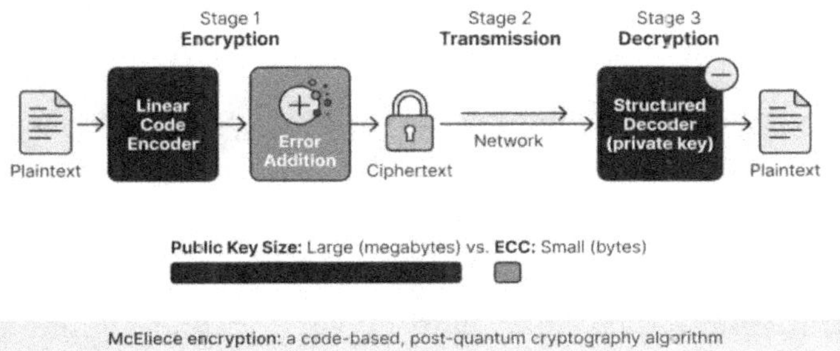

McEliece encryption: a code-based, post-quantum cryptography algorithm

3.4.2 Multivariate Polynomial Systems and Their Niche Applications

Multivariate polynomial cryptography builds on the difficulty of solving systems of multivariate polynomial equations over finite fields. Given a system of m quadratic equations in n variables over a small finite field, finding a solution is NP-hard in general. Multivariate schemes exploit this by constructing a public key as a set of such equations, while the private key contains a structured transformation that allows efficient solution. The security of these schemes has been less uniformly positive than that of lattice- or hash-based schemes: several multivariate proposals have been broken by algebraic attacks that exploit additional structure in the polynomial systems beyond what was originally analyzed.

NIST included MAYO and UOV (Unbalanced Oil and Vinegar) as additional signature candidates in its ongoing evaluation process. Multivariate signatures offer very fast verification — often faster than lattice-based signatures — and compact signature sizes in some variants. These properties make them attractive for specific applications: embedded systems with constrained resources, high-throughput certificate validation, and contexts where verification speed is a hard constraint, and signing is less frequent. The deployment case for multivariate schemes in a broad enterprise migration is limited by their less uniform security history and a less mature implementation ecosystem than lattice-based schemes. They are better positioned as niche supplements in a multi-algorithm strategy than as primary replacements for RSA or ECC.

3.4.3 Why Algorithm Diversity Is a Strategic Asset, Not Redundancy

Enterprise security architectures that depend on a single algorithm family for all cryptographic operations inherit all of that family's risk. If a mathematical advance breaks the hardness assumption underlying lattice-based cryptography — an unlikely but non-zero possibility that the cryptographic community takes seriously — then an enterprise that has migrated entirely to ML-KEM and ML-DSA has replicated the current situation: a single class of algorithms on which all security depends, now broken. A defense-in-depth approach to post-quantum migration uses different algorithm families for different use cases, where operationally feasible, so that a break in one family does not simultaneously compromise the entire cryptographic infrastructure.

The practical expression of algorithm diversity is a portfolio approach. Primary applications — TLS, PKI, authentication — migrate to NIST-standardized lattice-based schemes as the default. High-security, low-frequency signing operations — root CA signing, firmware signing — use hash-based schemes (SLH-DSA) for their conservative security foundation. Code-based alternatives are evaluated for use cases where key size is manageable, and security longevity is the dominant criterion. The portfolio is documented in a cryptographic architecture decision record, reviewed annually, and updated as new standards and implementation experience accumulate. This is not redundancy; it is responsible-by-design risk management that mirrors the defense-in-depth principles already embedded in mature security programs.

3.5 Isogeny-Based Cryptography: A Cautionary Lesson in Algorithm Evaluation

In 2022, the cryptographic community was sharply reminded that algorithmic confidence must be calibrated to the depth of public scrutiny, not to the elegance of mathematical construction. SIDH — the Supersingular Isogeny Diffie-Hellman key exchange protocol — was broken by a classical mathematical attack after having been a leading post-quantum key exchange candidate through multiple rounds of NIST evaluation. The attack used sophisticated algebraic-geometry tools to recover secret keys efficiently, in a way that no prior analysis had anticipated. The SIDH break is the most important case study in post-quantum algorithm evaluation that enterprise security leaders can study, because it directly illustrates the risks of deploying algorithms before the global research community has had sufficient time and motivation to probe them thoroughly.

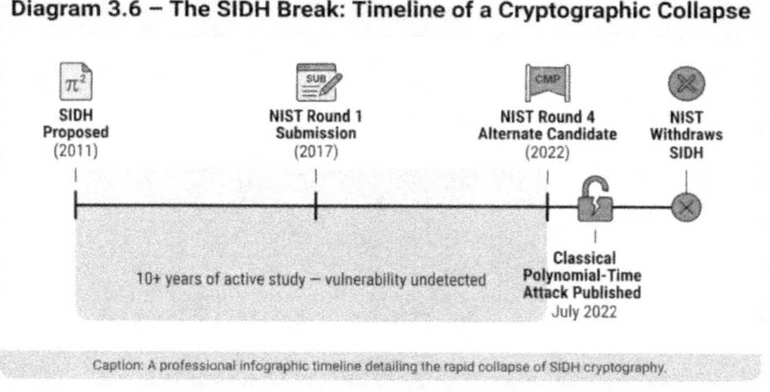

Diagram 3.6 – The SIDH Break: Timeline of a Cryptographic Collapse

Caption: A professional infographic timeline detailing the rapid collapse of SIDH cryptography.

3.5.1 The Rise and Fall of SIDH as a Reminder of Cryptographic Humility

SIDH was attractive to cryptographers for several reasons. It offered very small public key sizes — competitive with classical ECDH — at post-quantum security levels, addressing the key size disadvantage that was a concern for other post-quantum families. It was based on the mathematics of elliptic curve isogenies, a sophisticated and relatively new area of algebraic geometry that seemed to provide a fresh set of hard problems with no obvious connection to the algebraic structures that quantum computers exploit. The NIST evaluation process included multiple rounds of public analysis, and no attack on the core scheme was found for over a decade of study.

The 2022 attack showed that SIDH had additional structure — beyond what was known at the time of its design — that could be exploited using tools from algebraic curve theory. A small number of researchers published the attack, which was quickly confirmed by the broader community. Within days, the key exchange properties of SIDH were completely broken. Variants and derivative schemes were subsequently affected as well. The speed of the collapse — from publication to complete break in a matter of days, after years of scrutiny — illustrates a fundamental limitation of pre-deployment security evaluation: the research community cannot guarantee that no attack exists, only that no attack has been found within the time and effort applied.

3.5.2 What the SIDH Break Teaches About Vetting New Constructions

The SIDH break provides concrete lessons for enterprise security leaders evaluating algorithm families for post-quantum migration. The first lesson is that years of scrutiny without a published break is evidence of security, not proof. The strength of that evidence scales with the volume and quality of the scrutiny — who was looking, how intensively, and with what tools. SIDH received significant scrutiny, but the algebraic geometry tools used in the successful attack were not widely applied to it before 2022. The lesson for vendors and enterprise architects evaluating novel post-quantum proposals outside the NIST standardization process is to ask specifically: which mathematical communities have reviewed this, and have they included researchers from the specialized areas most relevant to its construction?

The second lesson is that the key size and performance advantages of newer, less-studied constructions should be treated with appropriate skepticism. SIDH's small key sizes were a genuine advantage, but they reflected a construction that proved fundamentally fragile. An algorithm family that achieves better performance than competitors through a novel mathematical approach is not necessarily offering a free lunch — it may be achieving that performance by using a mathematical structure that is less well-understood and therefore less well-defended. For enterprise migrations, the correct engineering posture is to prefer algorithms with conservative, well-studied security foundations and adequate performance, rather than those with exceptional performance and shallow security track records. NIST-standardized algorithms — ML-KEM, ML-DSA, SLH-DSA — reflect exactly this preference. They are not the most compact or the fastest possible post-quantum algorithms; they are the most thoroughly analyzed ones available.

3.6 Choosing Among Families: A Framework for Use-Case Alignment

The practical question for an enterprise security architect is not which algorithm family is theoretically best — it is which family to use for each specific use case in the organization's environment. This alignment exercise requires mapping the operational characteristics of each family to the requirements of each use case. The relevant dimensions are: key and signature size constraints, signing and verification speed requirements, signing frequency and state management complexity, deployment environment resource constraints, and regulatory or compliance requirements. The framework below structures this mapping for the most common enterprise cryptographic use cases.

3.6.1 Matching Algorithm Families to Key Exchange, Signatures, and Encryption Needs

Key exchange — the function of securely establishing a shared secret between two parties — is the most performance-sensitive and latency-critical of the three core cryptographic functions. In TLS, SSH, and other handshake-based protocols, key exchange occurs during connection establishment. Lattice-based key encapsulation through ML-KEM is the recommended choice for key exchange in virtually all enterprise contexts. Its performance is competitive with classical ECC, its implementation ecosystem is mature, and it has been standardized in FIPS 203. The key sizes, while larger than ECC, are well within the operational tolerances of modern network protocols. Hash-based and code-based schemes are not appropriate for key exchange — hash-based schemes produce only signatures, and code-based schemes with adequate security have prohibitively large key sizes for per-connection key exchange.

Digital signatures — used in certificates, code signing, software updates, and authenticated communications — have a wider range of appropriate algorithm choices, because signature operations are less latency-critical than key exchange in most enterprise contexts. ML-DSA is the recommended default for most signature use cases: it is standardized (FIPS 204), has reasonable signing and verification performance, and integrates with the same lattice-based ecosystem as ML-KEM. SLH-DSA is the recommended choice for high-security, low-frequency signing operations — particularly root CA certificate signing and firmware signing — where the hash-function-only security foundation is the dominant criterion and signature size is not operationally constraining. Multivariate signature schemes remain under evaluation for specialized high-throughput verification contexts and constrained devices.

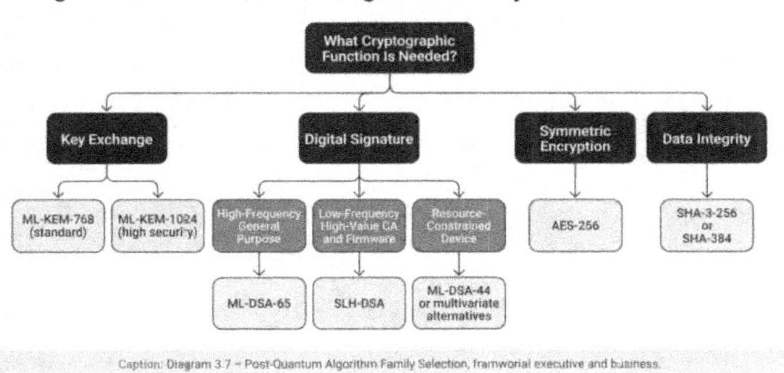

Diagram 3.7 – Post-Quantum Algorithm Family Selection Framework

Caption: Diagram 3.7 – Post-Quantum Algorithm Family Selection, framworial executive and business.

3.6.2 Performance, Key Size, and Operational Overhead: Building a Decision Matrix

Enterprise architecture teams benefit from a structured comparison of post-quantum algorithm families across the dimensions most relevant to deployment decisions. The key dimensions are: public key or encapsulation key size (affects certificate sizes, bandwidth, and PKI storage); ciphertext or signature size (affects network overhead per operation); signing and verification speed (affects throughput for high-frequency operations); implementation maturity (affects availability in cryptographic libraries and hardware acceleration); and security track record (affects confidence calibration and risk acceptance criteria). These dimensions create a multi-axis optimization problem that a single best answer cannot resolve — the correct choice depends on the specific requirements of the use case.

For the typical enterprise TLS migration — the highest-volume cryptographic use case in most organizations — the decision is effectively resolved by ML-KEM's performance and implementation maturity. For certificate issuance and PKI operations — which determine the trust anchor of the entire cryptographic infrastructure — the security track record dimension takes priority, supporting either ML-DSA or SLH-DSA depending on the CA system's signing frequency and state management capacity. For IoT and embedded systems, the key and signature size dimensions become constraining and require a detailed per-device evaluation. Documenting these decisions as cryptographic architecture decision records — with explicit rationale, applicable use cases, and review triggers — creates the operational clarity that sustains the migration program across personnel changes and technology refreshes.

3.6.3 When to Use Multiple Families for Defense-in-Depth

Using multiple algorithm families simultaneously — hybrid classical and post-quantum during the transition, and multiple post-quantum families after — adds operational complexity. It requires maintaining implementations of more algorithms, managing more key material types, and ensuring interoperability across a wider range of configurations. For these reasons, the temptation is to standardize on a single post-quantum family and deploy it universally. This temptation should be resisted for high-value use cases, even if it is accepted for lower-value bulk applications.

The defense-in-depth case for multiple families is strongest at the architecture layers that serve as trust anchors: PKI root keys, HSM master keys, and the cryptographic foundations of identity and authentication infrastructure. These are the layers where a single algorithm family break would have the most catastrophic enterprise-wide consequences. Using SLH-DSA for root CA signing keys alongside ML-DSA for end-entity certificates, for example, means that a break in ML-DSA does not immediately compromise the root of trust — the root was signed with a different, independently secure algorithm. This is not redundancy for its own sake; it is a deliberate architectural choice to prevent single-family failures from cascading through the entire cryptographic trust hierarchy. The pilot-to-production pathway for these high-value trust anchor applications should explicitly evaluate and incorporate multi-family diversity as a design criterion, not as an afterthought.

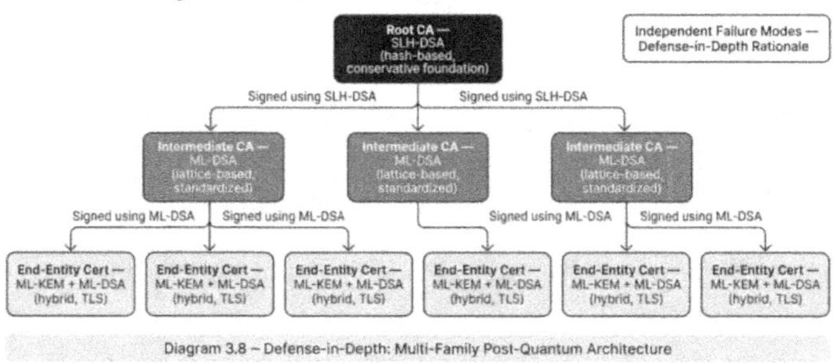

Diagram 3.8 – Defense-in-Depth: Multi-Family Post-Quantum Architecture

3.7 Manager's Checklist: Chapter 3

Diagram 3.9 - Post-Quantum Algorithm Family Quick Reference for Managers

Algorithm Family	Security Foundation	NIST Standard Status	Key/Signature Size	Performance	Best-Fit Use Cases
Lattice (ML-KEM, ML-DSA)	Hard lattice problems	Approved (FIPS 203, 204)		Moderate	General Purpose (Key Encapsulation & Signatures)
Hash-Based (SLH-DSA)	Hash function security	Approved (FIPS 205)		Slow	Firmware Updates, Critical Signatures (High Confidence)
Code-Based (Classic McEliece)	Hardness of decoding (Goppa codes)	Alternate/Round 4 Candidate		Fast	High-Security Key Exchange (Large Key Tolerance)
Multivariate (MAYO, UOV)	Hardness of quadratic systems over finite fields	Candidates (Under Review)		Fast	Short Digital Signatures (Resource-Constrained Devices)

Diagram 3.9 – Post-Quantum Algorithm Family Quick Reference for Managers

- Confirm that your post-quantum migration plan has defaulted to NIST-standardized algorithms — ML-KEM for key exchange, ML-DSA for signatures — and document the rationale for any departures from these defaults.

- Identify all root and intermediate CA signing operations in your PKI hierarchy and evaluate whether SLH-DSA is appropriate for root key signing, given its conservative hash-function-only security foundation.

- Inventory any IoT, embedded, or resource-constrained devices that participate in cryptographic operations and flag them for a separate algorithm selection analysis — lattice-based defaults may not fit all constrained environments.

- Confirm that your migration plan includes a cryptographic architecture decision record format that documents algorithm choices, applicable use cases, performance benchmarks, and review triggers.

- Assess whether your vendor ecosystem supports both ML-KEM and ML-DSA in their cryptographic libraries, hardware acceleration units, and PKI software — implementation maturity is a deployment prerequisite.

- Determine whether any high-value trust anchor keys in your organization — root CA, HSM master keys, firmware signing keys — should use a different algorithm family from the enterprise default as a defense-in-depth measure, and document that decision explicitly.

- Establish a monitoring mechanism for cryptographic research publications — specifically for mathematical advances affecting lattice problems, hash function properties, or code-based assumptions — so that the organization can respond rapidly if the security landscape for any deployed algorithm family changes.
- Review the SIDH break as a case study with your security architecture team and apply its lessons to any non-NIST-standardized post-quantum proposals you are being asked to evaluate or procure.

3.8 Hardness Built to Last

The post-quantum algorithm families described in this chapter are not perfect — no cryptographic system can claim that. They are the best available: mathematically grounded in problems that quantum computers cannot efficiently solve, publicly scrutinized through one of the most rigorous competitive evaluations in the history of cryptographic standardization, and now codified in NIST standards that enterprise architects, vendors, and regulators are building around.

The SIDH break is a reminder that cryptographic confidence must be earned through sustained scrutiny and not simply assumed from mathematical elegance. The algorithm families that survived the NIST process — lattice-based, hash-based, and code-based — did so because the global research community examined them thoroughly and found no shortcuts. That does not close the door permanently; cryptography is a living discipline, and the migration program you build must include the institutional capacity to respond if the landscape changes. But it does mean that the migration ahead is moving toward genuinely stronger foundations, not merely exchanging one set of assumptions for another of equivalent fragility.

Managers who understand why lattices resist quantum attacks, why hash-based schemes carry the most conservative security argument available, and why algorithm diversity is a risk management strategy — not just an engineering preference — are equipped to make decisions that are both technically defensible and operationally aligned with their organization's risk posture. The next chapter turns to the standards themselves: what NIST has finalized in FIPS 203, 204, and 205, and what those standards mean for the specific procurement, configuration, and compliance decisions your enterprise must now make.

4 The Standards Have Landed: Decoding the NIST Post-Quantum Selections and Their Enterprise Implications

The email arrives on a Tuesday morning. Your CISO has forwarded a memorandum from your federal compliance team. NIST has finalized three post-quantum cryptographic standards — FIPS 203, 204, and 205 — and the clock on your migration has officially started. You are the IT program manager responsible for a hybrid infrastructure spanning on-premises data centers, three cloud providers, dozens of SaaS platforms, and several hundred customer-facing APIs. You have a team of twelve engineers, a compliance officer who asks excellent questions you rarely have complete answers to, and a budget cycle that renews in October. Your CISO wants a briefing in two weeks on what the standards mean for your organization specifically and what you are going to do about them.

That scenario — or a version of it — is playing out in security and infrastructure teams across every sector. The standards are real, they are final, and the enterprise implications are concrete. This chapter decodes what was selected, why NIST selected it, what the documents themselves say that enterprise teams need to act on, and what the regulatory consequences are in federal, healthcare, and commercial environments. It also names the honest gaps the standards do not resolve, so your planning is not built on false certainty.

The NIST standardization of post-quantum algorithms is not a research milestone — it is an operational mandate. Every system your organization relies on that uses public-key cryptography for key exchange or digital signatures is now operating on an algorithm class that has a known termination date. The termination date is not yet fixed to a calendar year, but the mechanism of termination is understood: a cryptographically relevant quantum computer running Shor's algorithm. The standards are the beginning of the migration runway, not the end of the planning window.

The workflow-level impact is immediate in at least two directions. First, procurement decisions made today — hardware security modules, cloud KMS services, PKI infrastructure, VPN concentrators, TLS termination devices — must account for post-quantum readiness or they lock you into another migration within a few years. Second, any long-lived data encrypted today under RSA or ECC keys is a harvest-now-decrypt-later target. The standards give you the approved replacement tools. Whether your organization uses them in time is a program management question, not a cryptography question.

4.1 The NIST Post-Quantum Standardization Process: A Brief History

4.1.1 Why NIST Ran a Competitive Evaluation Rather Than Direct Selection

When NIST began its post-quantum standardization project in 2016, it faced a problem that had no precedent in modern cryptographic standardization. The threat — a large-scale quantum computer running Shor's algorithm — was theoretically proven but had not been built in practice. The candidate algorithms were mathematically novel, based on problems that had not been stress-tested by decades of cryptanalytic work, the way RSA and elliptic curve cryptography had been. Selecting a single algorithm by committee consensus would have produced a standard with unknown confidence bounds. A competitive evaluation, open to global cryptographic researchers, was the only method credible enough to justify the resulting standards.

The decision to make the competition open was strategically important. By inviting academic teams, national security researchers, and commercial cryptographers from around the world to submit, analyze, and attack candidate algorithms, NIST harnessed a global cryptanalytic workforce far larger than any single organization could assemble. The attacks that emerged during the competition — including the 2022 break of SIDH/SIKE — validated the process design. An algorithm that would have been deployed at scale was removed before finalization. That outcome represents the process working as intended.

Enterprise managers should understand that the competitive process was deliberately conservative. Candidates that survived eight years of public analysis carry a confidence level that no internally developed or proprietary algorithm can match. That conservatism is not bureaucratic caution — it is mission-aligned risk management at the global scale of digital infrastructure.

Diagram 4.1 – NIST PQC Standardization Timeline

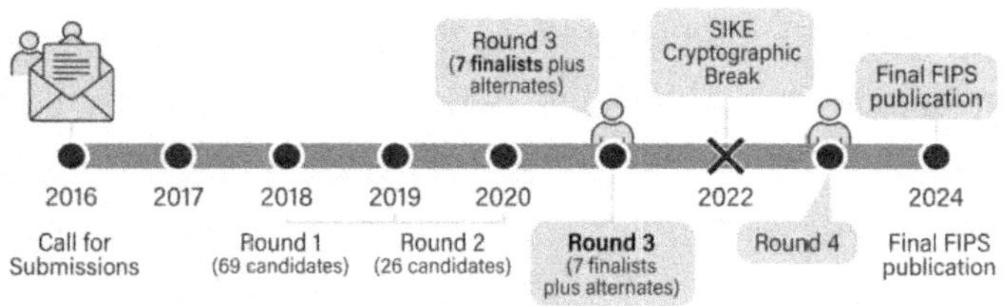

4.1.2 The Submission Rounds and How Candidates Were Narrowed

The competition opened with sixty-nine complete submissions in 2017. Over four evaluation rounds, NIST and the global cryptographic community subjected each candidate to public scrutiny on four dimensions: security against both classical and quantum attacks, performance across constrained and high-throughput environments, implementation characteristics, and flexibility across use cases. The evaluation was not a beauty contest. Algorithms were eliminated because of cryptanalytic breaks, because of implementation vulnerabilities, because of performance characteristics that made them unsuitable for real-world deployment, and because they did not offer sufficiently distinct properties to justify standardizing multiple algorithms within the same family.

By Round 3, the field had narrowed to seven finalists and eight alternates. The finalists were grouped by use case: key encapsulation mechanisms (KEMs) for confidentiality, and digital signatures for authentication and integrity. The alternates represented either different mathematical families that provided hedging value if a finalist family proved vulnerable, or algorithms with specialized performance profiles suited to constrained environments.

The four rounds took from 2017 to 2022 to complete. This timeline reflects the genuine difficulty of the evaluation work, not organizational inefficiency. The cryptanalytic depth required to develop confidence in novel mathematical constructions takes years of sustained academic attention. Enterprise managers who are impatient with the timeline should weigh that impatience against what would have happened if SIKE had been standardized and deployed at scale before its classical break was discovered.

4.1.3 The Role of External Cryptanalysis in Building Standards Confidence

Standards confidence in post-quantum cryptography is built through published attacks. Every algorithm that entered the competition was subjected to cryptanalytic attempts by teams that had professional and reputational incentives to succeed. A published break that eliminates a candidate increases confidence in the surviving candidates because it demonstrates that the evaluation process has teeth. An algorithm that survives concentrated global attack for eight years carries an earned confidence level.

The external cryptanalysis also produced refinements. Candidate algorithms were revised between rounds in response to attack findings that did not constitute full breaks but revealed structural weaknesses. Those revisions were themselves subjected to additional scrutiny. The algorithms that reached finalization are therefore not the algorithms that were submitted in 2017 — they are hardened versions that incorporated lessons from years of adversarial review.

Enterprise architects should understand that the parameter sets in the final standards reflect specific, analyzed security levels, not arbitrary choices.

Ongoing cryptanalysis does not stop at finalization. The security community will continue to examine the selected algorithms. A mature enterprise cryptographic governance posture treats algorithm monitoring as a continuous practice, not a one-time project. The standards documents themselves acknowledge that parameter sets may need revision if new analysis reduces confidence in specific security levels.

4.2 FIPS 203, 204, and 205: What Was Selected and Why

4.2.1 ML-KEM (CRYSTALS-Kyber): The Key Encapsulation Standard

FIPS 203 standardizes ML-KEM, the Module Lattice-based Key Encapsulation Mechanism derived from the CRYSTALS-Kyber algorithm. A key encapsulation mechanism is the building block that enables two parties to establish a shared secret over an untrusted channel — the function that makes TLS, VPNs, and encrypted messaging systems work. ML-KEM replaces the role played by RSA key transport and Diffie-Hellman key exchange in current protocols.

The security of ML-KEM rests on the hardness of the Module Learning With Errors problem, a lattice-based construction that is believed to resist both classical and quantum attacks. The underlying mathematical structure involves noisy linear algebra over polynomial rings — a problem space that has been studied since the early 2000s and for which no quantum algorithm providing meaningful speedup is currently known. The word "believed" in the previous sentence is not a hedge inserted for legal caution. It is a factual statement about the current state of cryptanalytic knowledge, and managers should carry that epistemic honesty into their planning.

FIPS 203 defines three parameter sets corresponding to different security levels: ML-KEM-512 (roughly equivalent to AES-128 security), ML-KEM-768 (roughly equivalent to AES-192 security), and ML-KEM-1024 (roughly equivalent to AES-256 security). For most enterprise use cases involving TLS and key exchange, ML-KEM-768 represents the practical default — it provides a security level that exceeds the requirements of most compliance frameworks while maintaining performance characteristics acceptable for high-throughput environments. ML-KEM-1024 is appropriate for long-lived keys, high-value credential management, and environments with explicit regulatory requirements for higher security margins.

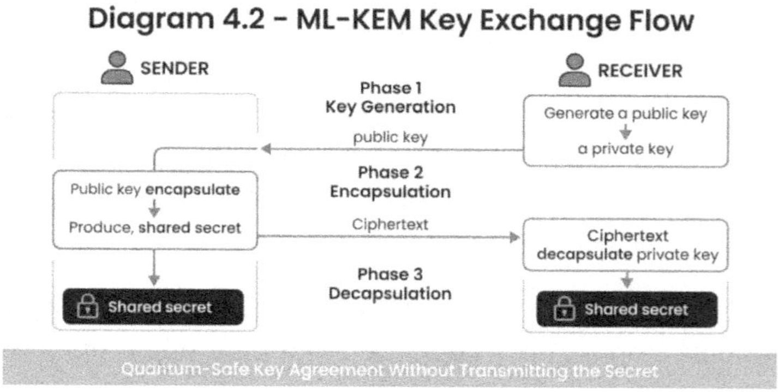

Diagram 4.2 – ML-KEM Key Exchange Flow

Key size implications are practical characteristics that enterprise architects must plan for. An ML-KEM-768 public key is approximately 1,184 bytes; the encapsulation ciphertext is approximately 1,088 bytes. By comparison, a 256-bit ECDH public key is 32 bytes. This size increase does not render ML-KEM impractical, but it requires attention in systems with constrained message sizes, tight latency requirements, or bandwidth-sensitive protocols. Applications that embed key material in QR codes, RFID-adjacent systems, or legacy protocol stacks with hardcoded field sizes will require more than a simple algorithm swap.

4.2.2 **ML-DSA (CRYSTALS-Dilithium): The Digital Signature Standard**

FIPS 204 standardizes ML-DSA, the Module Lattice-based Digital Signature Algorithm derived from CRYSTALS-Dilithium. Digital signatures are the mechanism that proves a message, document, certificate, or software artifact was produced by a specific entity and has not been tampered with. They underpin code signing, certificate authorities, document authentication, email security, and the PKI infrastructure that connects every secure enterprise system.

ML-DSA uses the same Module Learning With Errors hardness foundation as ML-KEM, which is both a strength and a point of concentration. The strength is that lattice-based security has received sustained academic attention and has not yielded to quantum attack strategies. The risk concentration is that if a future mathematical breakthrough undermined the Module LWE problem, both ML-KEM and ML-DSA would be affected simultaneously. Enterprise architects building defense-in-depth strategies should note this correlation and consider SLH-DSA as a signature hedge.

FIPS 204 defines three parameter sets: ML-DSA-44, ML-DSA-65, and ML-DSA-87, corresponding to security levels comparable to 128, 192, and 256 bits, respectively. Signature sizes range from approximately 2,420 bytes at the ML-DSA-44 level to approximately 4,627 bytes at the ML-DSA-87 level. For context, an ECDSA signature over P-256 is typically 64 bytes. This size increase has material implications for any system that processes high volumes of signatures — log signing, transaction authentication, certificate validation chains, and code signing pipelines — and must be factored into performance architecture before migration.

Diagram 4.3 – ML-DSA vs. Classical Signature Size Comparison

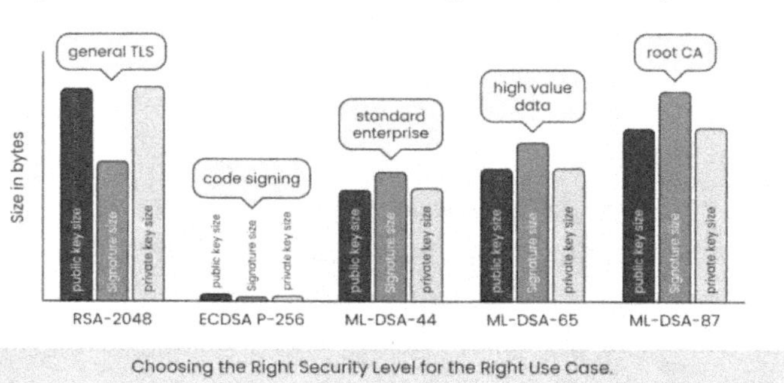

Choosing the Right Security Level for the Right Use Case.

4.2.3 SLH-DSA (SPHINCS+): The Stateless Hash-Based Signature Standard

FIPS 205 standardizes SLH-DSA, the Stateless Hash-Based Digital Signature algorithm derived from SPHINCS+. Unlike ML-DSA, which relies on lattice-based hardness, SLH-DSA derives its security entirely from the properties of cryptographic hash functions. This makes SLH-DSA the most conservative of the three standards in terms of security assumptions. If hash functions remain secure (and SHA-2 and SHA-3 have decades of confidence behind them), SLH-DSA remains secure regardless of what happens to lattice mathematics.

The conservatism of SLH-DSA comes with performance costs that enterprise teams must evaluate candidly. SLH-DSA signatures are significantly larger than ML-DSA signatures — ranging from approximately 7,856 bytes for the smallest parameter set to over 49,000 bytes for the largest. Signing operations are also substantially slower than lattice-based alternatives. These characteristics make SLH-DSA unsuitable as a general-purpose replacement for ECDSA in latency-sensitive applications.

Where SLH-DSA delivers genuine value is in use cases where the conservatism of the security assumption justifies the performance cost. Root certificate authorities, firmware signing for critical infrastructure, long-lived code signing certificates, and high-value document authentication are contexts where the absence of structural assumptions beyond hash function security is a meaningful advantage. Enterprises deploying both ML-DSA and SLH-DSA for different use cases are not duplicating effort — they are implementing algorithm diversity as a risk management strategy. If lattice-based cryptography were ever broken, SLH-DSA deployments would remain secure.

NIST PQC Standards Portfolio Overview

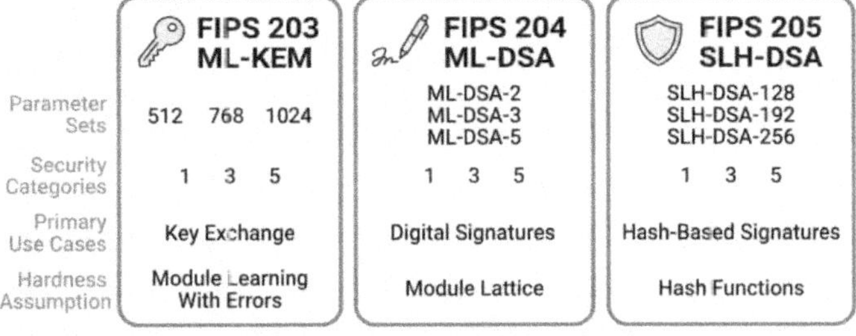

4.2.4 **FN-DSA (FALCON): The Fourth Signature Candidate and Its Status**

FALCON, standardized as FN-DSA in a separate FIPS publication, uses NTRU lattice constructions rather than the Module LWE problem. FALCON produces significantly smaller signatures than ML-DSA — comparable to ECDSA sizes in many configurations — and has strong performance characteristics. Its inclusion in the standardization portfolio reflects the value of having a compact lattice-based signature option alongside the more conservative ML-DSA.

Enterprise adoption of FN-DSA is more cautious than that of ML-KEM and ML-DSA, primarily because constant-time implementation is genuinely challenging. FALCON's signing algorithm involves Gaussian sampling over lattices, an operation that is notoriously susceptible to timing side-channels if not implemented with explicit care. Commercial implementations that have undergone rigorous side-channel validation are still emerging. For most enterprise environments, ML-DSA is the operationally safer choice today, with FN-DSA as a long-term option for environments where signature compactness is a hard constraint and implementation quality can be verified.

4.3 **Reading the Standards Documents: What Enterprise Teams Need to Extract**

4.3.1 **Parameter Sets, Security Levels, and How to Match Them to Use Cases**

Each of the three finalized FIPS standards defines multiple parameter sets corresponding to classical security equivalents. The mapping is expressed in terms of security categories aligned to AES key lengths. Category 1 is broadly equivalent to breaking AES-128; Category 3 to AES-192; Category 5 to AES-256. These categories describe the minimum effort required to mount a successful attack against the defined parameter set, not the effort required to break the underlying primitive.

Matching parameter sets to use cases requires a decision framework rather than a single answer. The relevant dimensions are: the sensitivity of the data or system being protected, the expected lifetime of the key material or signed artifact, the application's performance requirements, and the compliance floor established by applicable regulations. For most enterprise applications in the current environment, Category 3 parameter sets — ML-KEM-768, ML-DSA-65, and SLH-DSA with 192-bit-equivalent parameters — represent a reasonable default that exceeds most regulatory minimums while maintaining acceptable performance.

The decision becomes more nuanced for specific contexts. Long-lived root CA keys, master key encryption keys for hardware security modules, and keys protecting data with confidentiality requirements extending decades into the future warrant Category 5 parameters. The signing keys for everyday TLS certificates and short-lived authentication tokens can operate effectively at Category 1 or Category 3. Matching parameters to use cases is not a one-size-fits-all exercise, and the standards documents provide the technical foundation for making those distinctions appropriately.

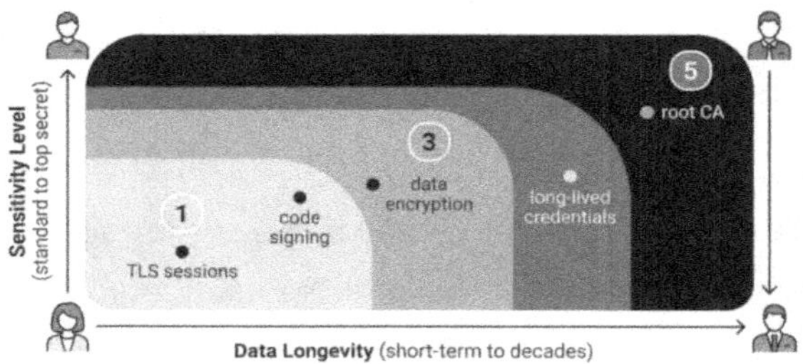

4.3.2 Interoperability Requirements and Protocol Integration Guidance

The FIPS standards documents address the cryptographic primitives in detail but are deliberately silent on protocol-specific integration. Protocol-level guidance for how to embed ML-KEM into TLS 1.3, how to structure hybrid certificates for X.509 PKI, and how to negotiate post-quantum key exchange in SSH is produced by separate standards bodies — primarily the IETF — working in parallel with the NIST process. Enterprise architects must track both because a standards-compliant ML-KEM implementation that is incorrectly integrated into a TLS handshake does not provide the intended security.

The practical interoperability work happening as of the final FIPS publications includes IETF RFCs for post-quantum key exchange in TLS 1.3 (X25519MLKEM768 and related hybrid groups), certificate profile extensions for post-quantum public keys and signatures, and SSH key exchange method identifiers for quantum-resistant algorithms. Libraries, including OpenSSL, BoringSSL, and the major cloud provider TLS implementations, began shipping post-quantum integration in experimental and then production modes starting in 2023 and 2024. Procurement teams should verify whether specific library versions support these integrations and whether the implementations have been formally evaluated.

4.3.3 What the Standards Say About Migration Timelines and Deprecation

The FIPS standards documents do not specify mandatory migration deadlines. They define the algorithms, their parameter sets, and their operational characteristics. The migration timelines appear in complementary guidance from regulatory bodies — CISA, NSA, OMB, and FIPS 140 validation requirements — rather than in the algorithm standards themselves. This distinction matters for enterprise planning because it means the urgency timeline is not universally derived from the same source document. Federal agencies, defense contractors, healthcare organizations, and commercial enterprises face different externally imposed timelines even though they are all working toward the same algorithmic destination.

What the standards do establish is a clear signal for the deprecation of classical algorithms. Federal agencies are directed by existing guidance to deprecate RSA, ECDSA, and classical Diffie-Hellman for key exchange and signature applications within specific planning horizons, which have been progressively tightened as the standards have been finalized. Commercial organizations without direct federal mandates should treat those federal timelines as proxies for industry direction — the insurance, financial services, and healthcare regulatory environments are already developing aligned guidance. Organizations that wait for their specific sector regulator to publish a hard deadline before beginning migration will find themselves compressing multi-year programs into windows that cannot accommodate that work safely.

4.4 Regulatory and Compliance Ripple Effects

4.4.1 Federal Mandates: NSA, CISA, OMB Guidance, and CNSS Policy 15

Federal agencies and contractors working in national security contexts face the most structured migration timelines in the current regulatory landscape. NSA's Commercial National Security Algorithm Suite 2.0 (CNSA 2.0) specifies the post-quantum algorithms approved for use in national security systems and establishes timelines for deprecating classical equivalents. The timelines are differentiated by system criticality and algorithm type, with software and firmware signature updates expected earlier than full key exchange migrations. Enterprises with any nexus to national security systems — whether as prime contractors, subcontractors, or cloud service providers handling regulated data — should read CNSA 2.0 as a hard schedule rather than general guidance.

CISA's post-quantum guidance documents and the Office of Management and Budget's memoranda to federal agencies provide the broader framework for the civilian federal migration. The CNSS Policy 15 documents establish the policy architecture for national security information systems and have historically preceded broader federal standards by two to three years, making them an early-warning indicator of where civilian agency requirements will move. Organizations that have previously treated NSA guidance as irrelevant to their commercial operations should reconsider that posture. The encryption standards that underpin commercial infrastructure do not exist in a separate regulatory universe from national security requirements.

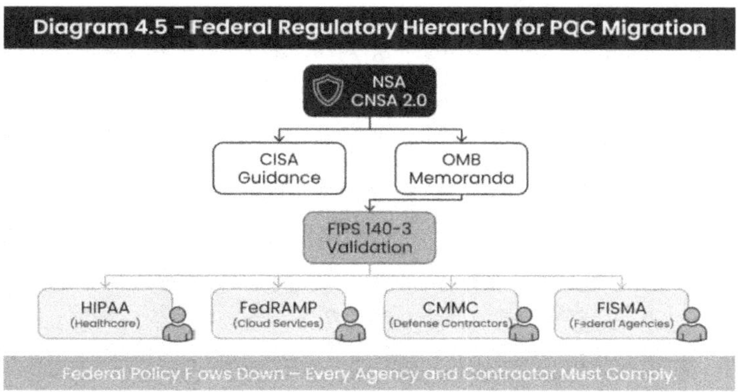

4.4.2 Healthcare Compliance Implications Under HIPAA and HITECH

Healthcare organizations face a distinct compliance challenge in the post-quantum era because the Protected Health Information they encrypt today may remain sensitive for decades. A patient's genetic information, HIV status, or mental health history does not expire. Harvest-now-decrypt-later attacks against healthcare data are not a theoretical future threat — they are an active strategic interest for foreign intelligence services that value long-horizon health data for coercion, blackmail, and demographic targeting purposes.

HIPAA's Security Rule does not specify cryptographic algorithms by name — it requires "addressable" implementation specifications for encryption of data in transit and at rest, with the reasonableness standard set by current industry practice. As post-quantum standards become finalized and regulatory guidance evolves, organizations that continue to rely exclusively on RSA-2048 or ECDSA for health data protection will face increasing difficulty demonstrating that their encryption posture meets the "reasonable and appropriate" standard. HHS Office for Civil Rights guidance and HITECH enforcement interpretations have historically followed NIST and NSA cryptographic guidance as proxies for industry best practice.

Healthcare organizations should note the particular sensitivity of systems that combine long data retention requirements with network-accessible interfaces. Electronic health record systems, clinical data repositories, genomics platforms, and payer data warehouses all fit this profile. The remediation priority framework for these organizations should prioritize harvest-now risk and begin a hybrid migration for data-in-transit protection before the full migration program is complete.

4.4.3 How FIPS 140-3 Validation Will Apply to Post-Quantum Implementations

FIPS 140-3 validation is the federal government's product certification program for cryptographic modules used in regulated environments. Federal agencies must use FIPS 140-3 validated cryptographic modules for sensitive but unclassified data, and many commercial and healthcare compliance frameworks reference FIPS 140-3 validation as a procurement requirement. The validation program tests whether a specific product implementation correctly and securely implements the claimed cryptographic algorithms — it is not a theoretical assessment but a product-specific evaluation.

Post-quantum algorithm support under FIPS 140-3 is still maturing as of the final FIPS publications. The CMVP (Cryptographic Module Validation Program) has published transition guidance establishing that modules will be validated against FIPS 203, 204, and 205 under the existing FIPS 140-3 framework. Vendors must submit their implementations for testing by accredited Cryptographic Testing Laboratories. The testing backlog and laboratory capacity constraints mean that validated post-quantum implementations will reach the market on a timeline that lags the standards themselves by months to years.

This gap creates a practical challenge for organizations with hard FIPS 140-3 procurement requirements. The options are: implement hybrid configurations that use validated classical algorithms alongside unvalidated post-quantum implementations (accepting that the classical component provides the validated protection while the post-quantum component adds forward defense); delay post-quantum migration until validated modules are available (accepting harvest-now risk during the delay); or implement post-quantum in environments where FIPS 140-3 is not a hard requirement as a capability-building measure while waiting for validated options in regulated contexts. Each option is a legitimate risk-management decision, and the right choice depends on your specific regulatory environment and data-sensitivity profile.

4.5 What the Standards Do Not Resolve: Open Questions for Enterprise Architects

4.5.1 Algorithm Flexibility and the Absence of a Universal Transition Deadline

The finalization of FIPS 203, 204, and 205 establishes what to migrate to. It does not establish a universal deadline for completing that migration. Federal agencies face sector-specific deadlines in NSA and OMB guidance. Commercial organizations face deadlines that vary by sector, by the nature of the data they handle, and by their contractual relationships with federal agencies or regulated entities. For many organizations, the honest answer to "when do we have to be done?" is "we don't know yet, but we know the direction, and we know the risk of delay." Operational clarity demands that this uncertainty be managed explicitly — through risk-based prioritization, staged migration plans, and stakeholder communications that are honest about the absence of a single universal deadline.

The absence of a hard, universal deadline is itself a risk-management challenge. Organizations that treat the absence of a deadline as permission to delay will find themselves in an accelerated migration posture when their sector regulator eventually publishes requirements. The organizations that use the current window to build migration infrastructure, cryptographic agility, and institutional expertise will be able to respond to regulatory acceleration without crisis-mode execution. That advantage is a measurable program outcome even if it does not appear on a compliance scorecard today.

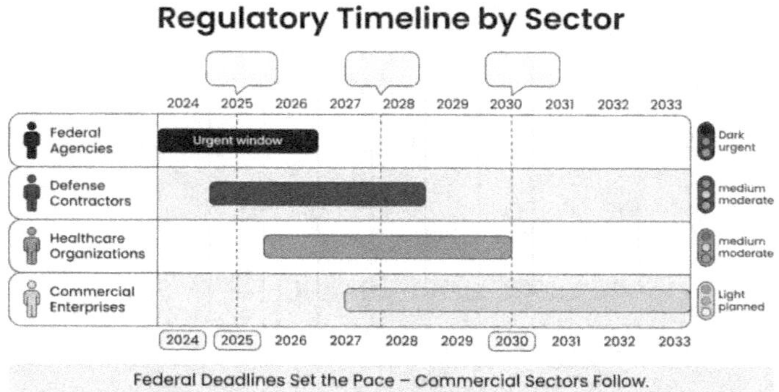

Federal Deadlines Set the Pace – Commercial Sectors Follow.

4.5.2 **Vendor Certification Timelines and the Gap Between Standards and Products**

A standard that exists on paper but has not been implemented in certified products creates a migration planning gap that enterprise architects must account for. The gap exists in multiple layers simultaneously. Cryptographic library support must precede operating system integration. Operating system integration must precede application-level support. Application-level support must precede enterprise procurement and deployment. And formal FIPS 140-3 validation of the complete implementation must precede deployment in regulated environments.

The practical consequence is that even organizations with full organizational commitment to post-quantum migration will face windows where the certified products they need do not yet exist. Hardware security modules with support for post-quantum algorithms are still rolling out from major vendors. PKI platforms with native post-quantum certificate authority support are in various stages of beta availability. In many cases, VPN concentrators and firewall platforms with ML-KEM integration are on vendor roadmaps rather than in general availability product catalogs.

Procurement teams should build a vendor roadmap assessment into their cryptographic migration planning now. Identifying which current vendors have published credible post-quantum roadmaps with specific delivery dates, which vendors are silent on the topic, and which vendors have shipped early implementations that require validation is a critical input to migration sequencing. Vendor selection and renewal decisions made without this information risk locking your organization into platforms that will extend your migration timeline rather than accelerating it.

Vendor Product Readiness Landscape

Product	Roadmap Announced	Beta Available	Generally Available	FIPS 140-3 Validated
HSMs			●	●
PKI Platforms		●	●	●
TLS Libraries		●	●	●
VPN Devices		●	●	●
Cloud KMS			●	●
Code Signing Tools			●	●

Legend
● FIPS validated
● Beta
Roadmap only
Not started

Know Where Your Vendors Stand Before Committing to a Migration Window

4.5.3 Jurisdictional Variation: Non-US Regulatory Bodies and Their Trajectories

NIST's post-quantum standards carry significant weight globally because of the deep integration of US cryptographic standards into global technology infrastructure. TLS, X.509 certificates, SSH, and the protocol ecosystem built on NIST-aligned algorithms are used in data centers and enterprises worldwide. The finalization of FIPS 203, 204, and 205, therefore, creates migration pressure in international organizations even in the absence of direct US regulatory jurisdiction.

Non-US standards bodies are developing their own post-quantum positions. The European Union Agency for Cybersecurity (ENISA) has published post-quantum migration guidance that is broadly aligned with NIST recommendations while acknowledging European-specific regulatory contexts. BSI in Germany, ANSSI in France, and the UK National Cyber Security Center have each published technical assessments and preliminary migration guidance. These documents are broadly consistent with NIST's recommendations for the ML-KEM and ML-DSA family of algorithms. However, some jurisdictions have expressed interest in additional algorithms not in the NIST portfolio and may standardize them separately.

Global organizations must therefore manage migration strategies that accommodate potential divergence between US and non-US requirements. The practical risk is not that different jurisdictions select fundamentally incompatible algorithms — the probability of that outcome is low. The risk is that timing requirements, specific parameter-set mandates, and certification program requirements vary enough across jurisdictions to create compliance complexity for organizations operating under multiple regulatory frameworks simultaneously. Building cryptographic agility into your architecture from the beginning is the structural answer to this complexity.

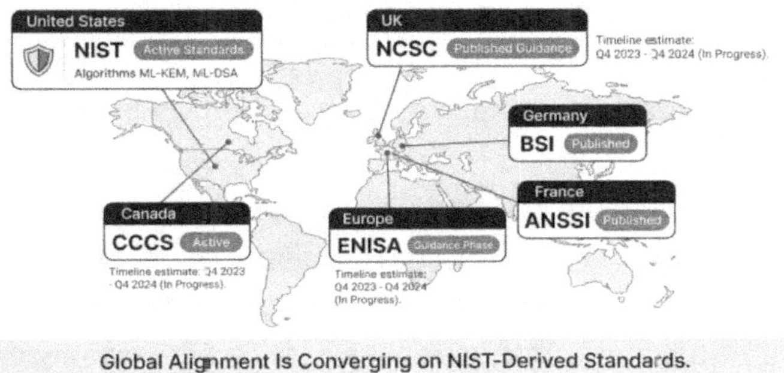

Global Alignment Is Converging on NIST-Derived Standards.

4.6 Manager's Checklist: Acting on the NIST Standards

- Confirm that your legal and compliance teams have reviewed CNSA 2.0, CISA's post-quantum guidance, and any sector-specific regulatory publications relevant to your organization.

- Build a parameter set decision matrix that maps your organization's use cases to the appropriate FIPS 203/204/205 security categories before any implementation decisions are made.
- Inventory which of your current technology vendors have published post-quantum roadmaps with specific delivery dates; escalate to procurement for vendors with no published position.
- Engage your PKI team to assess the certificate lifecycle and authority infrastructure implications of ML-DSA and SLH-DSA adoption, including root CA key replacement planning.
- Identify all systems in your environment that handle Protected Health Information or other long-retention sensitive data and flag them as harvest-now priority targets for early hybrid migration.
- Verify whether your FIPS 140-3 procurement requirements apply to post-quantum implementations in your regulated environments and determine the current validation status of candidate products.
- Establish a working relationship with your key TLS library and cryptographic toolkit vendors to stay current on post-quantum integration release schedules and known implementation issues.
- Draft a regulatory exposure briefing for your CISO that maps your compliance obligations by sector, identifies the most urgent deadlines, and distinguishes between hard mandates and recommended timelines.
- Document the algorithm selection rationale for any post-quantum cryptographic decisions made in the current

planning period — these records become the foundation for future audit evidence.

4.7 The Standards Are the Starting Line, Not the Finish Line

FIPS 203, 204, and 205 give enterprise teams the approved toolset for building quantum-resistant infrastructure. They do not hand you a migration plan, a vendor-certified product catalog, or a compliance deadline specific to your sector. What they give you is operational clarity on the direction: these are the algorithms, these are the security levels, and this is the mathematical foundation your future cryptographic architecture will rest on.

The enterprise implications flow from that clarity. Procurement decisions should now be made with post-quantum readiness as a selection criterion. Architecture decisions should be made with cryptographic agility as a design requirement. Risk decisions should reflect the harvest-now threat to long-lived sensitive data. And compliance planning should acknowledge that the timeline pressure is real even for organizations that have not yet received a sector-specific mandate, because the mandate will come and the preparation window will not expand to meet an unprepared organization.

The organizations that will navigate this transition with the least disruption are not necessarily the ones with the largest security teams or the biggest budgets. They are the ones that started building the foundation — cryptographic visibility, algorithm agility, vendor readiness assessment, and staged migration capability — before the external pressure forced the conversation. The standards have landed. The work begins here.

5 The Cryptographic Census: Finding Every Key, Certificate, and Algorithm Your Organization Depends On

Your organization has just been told by its external auditor that it needs to demonstrate cryptographic inventory coverage as part of its annual security attestation. The auditor wants to know which algorithms are in use, where key material lives, which certificates are approaching expiration, and which systems would be affected by a forced algorithm migration. You pull up your CMDB and your vulnerability scanner outputs. Neither contains what you need. The CMDB tracks hardware and software assets, not cryptographic properties. The vulnerability scanner flags known CVEs but does not enumerate algorithm usage across the application portfolio. Your network team has some TLS scan data from six months ago. Your development team has vague awareness that several applications use cryptographic libraries, but no systematic documentation of which ones or how they are configured.

This is not an unusual situation. It is the default situation in most organizations that have not run a deliberate cryptographic discovery program. The gap between "we have security tools" and "we have cryptographic visibility" is larger than most teams expect. It surfaces at the worst possible moments — during audits, during incident response, and during the planning stages of a migration that reveals dependencies no one mapped. This chapter provides the methodology for closing that gap: how to discover every key, certificate, and algorithm your organization depends on, how to turn raw discovery output into a risk-stratified register, and how to sustain that visibility as a continuous practice rather than a one-time project.

The case for cryptographic visibility is not abstract. It is grounded in the specific, practical problem of not knowing what you are protecting and what your protection mechanisms are. Organizations that cannot enumerate their cryptographic assets face four concrete risks that compound each other. First, they cannot accurately assess the scope of their quantum migration, leading to plans that are either over-resourced for low-impact systems or catastrophically under-resourced for high-impact ones. Second, they are exposed to certificate-expiration incidents when certificates are renewed without visibility into all the environments they are deployed in. Third, they cannot demonstrate compliance to auditors and regulators, who are increasingly asking algorithm-level questions. Fourth, they cannot perform vendor risk assessments that account for cryptographic dependencies in the supply chain.

The workflow-level impact of poor cryptographic visibility manifests during migration execution. Teams that begin post-quantum migration without a complete inventory routinely discover unexpected dependencies mid-migration — a legacy API using an EC key pair for client authentication. This third-party library implements its own TLS stack and does not inherit system-level configuration, nor does it support a hardware device with firmware-embedded certificate authorities that cannot be updated remotely. Each of those discoveries adds scope, delays timelines, and forces re-sequencing of dependencies. The cryptographic census is not overhead on the migration program. It is the foundation for every migration decision.

A complete cryptographic census is also a living security artifact with value beyond post-quantum migration. It reveals shadow cryptography — encryption and signing operations that occur outside approved channels and algorithms. It reveals hardcoded credentials and embedded keys that pose immediate security risks, independent of quantum threats. And it provides the evidence base that audit teams and regulators need to assess your cryptographic posture with specificity rather than in a general way.

5.1 Why Standard Asset Inventories Miss the Cryptographic Dimension

5.1.1 The Gap Between IT Asset Management and Cryptographic Visibility

IT asset management systems are designed to track the existence and configuration of hardware and software assets. They record hostnames, IP addresses, operating system versions, software inventories, and patch levels. They do not, by design, record which cryptographic algorithms an application uses when it establishes a TLS connection, which key lengths are configured on a certificate authority, or whether a software package links against a vulnerable version of a cryptographic library. The gap is not a deficiency in your CMDB — it reflects the fact that cryptographic properties are attributes of how software operates, not how it is inventoried.

The same limitation applies to configuration management databases, software asset management tools, and most vulnerability management platforms. A vulnerability scanner can tell you that a system is running OpenSSL 1.1.1 and that this version is affected by a specific CVE. It cannot tell you that the application calling OpenSSL is using RSA-2048 for key exchange, that the key was generated four years ago and will need replacement under your post-quantum migration plan, and that the same certificate is deployed on fourteen other hosts using the same serial number. That chain of information requires a deliberate cryptographic discovery methodology.

The implications for managers are direct: do not assume that your existing asset management investments have produced cryptographic visibility. They have not, even if they are mature and well-maintained. Cryptographic visibility requires a separate, dedicated discovery effort that combines network scanning, application analysis, certificate inventory, and hardware audit using tools and methods specifically designed for cryptographic property discovery.

Diagram 5.1 – The Cryptographic Visibility Gap

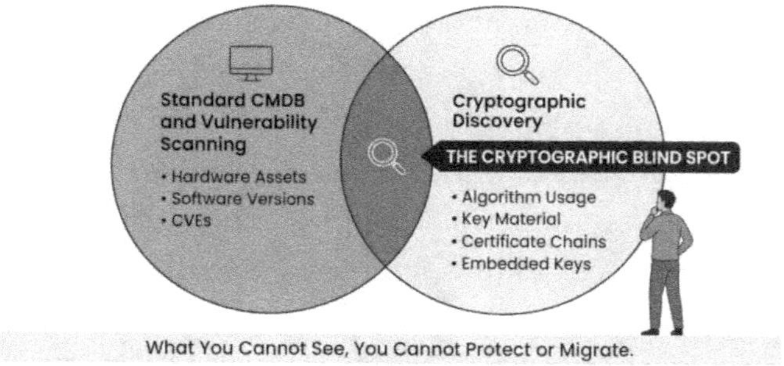

What You Cannot See, You Cannot Protect or Migrate.

5.1.2 Hardcoded Keys, Embedded Certificates, and the Shadow Cryptography Problem

Shadow cryptography refers to cryptographic operations that occur outside of the organization's managed cryptographic infrastructure — operations that are not tracked in PKI systems, not monitored by security tools, and not covered by key management policies. Shadow cryptography is pervasive in most mature enterprise environments because it accumulates silently over years of development and procurement decisions made without cryptographic visibility requirements.

The most common manifestations include hardcoded cryptographic keys embedded in application source code or configuration files, self-signed certificates generated by individual developers for testing environments that migrated into production, API keys and HMAC secrets distributed across multiple systems without a central inventory, and vendor-supplied devices with factory-installed keys that are never rotated. Each of these represents a cryptographic asset that will not appear in a standard security scan, will not be included in certificate expiration monitoring, and will not be on the radar of a post-quantum migration team working from standard inventory sources.

Discovering shadow cryptography requires active code analysis, configuration audit, and behavioral monitoring — approaches that are inherently more invasive and time-consuming than passive network scanning. The discovery methodology section of this chapter addresses the tooling and techniques for surfacing these assets. Still, the first step is organizational: acknowledging their presence in quantity and building the programmatic will to find them. In most enterprise environments, the shadow cryptography discovery phase produces the largest and most surprising portion of the cryptographic inventory.

5.1.3 **Third-Party Libraries and the Transitive Dependency Trap**

Modern enterprise applications depend on cryptographic functionality that they do not implement directly. A typical web application's TLS stack, key generation, certificate validation, and hashing functions are implemented by a chain of libraries — the application framework calls a higher-level crypto library, which in turn calls a lower-level primitive library, which in turn calls the operating system's cryptographic services. This chain of transitive dependencies means that understanding your cryptographic surface area requires tracing not just what your code does directly but what every library your code depends on does, and what every library those libraries depend on does.

The transitive dependency problem is particularly acute for organizations that have adopted microservices architectures, polyglot development environments, or frequent library updates. A single update to a third-party dependency can change the cryptographic algorithms in use without any developer intentionally making a cryptographic decision. An npm package update that pulls in a new version of a node-forge dependency, a Python requirements.txt that pins cryptography to a version with different default cipher suites, or a Go module update that changes the default key agreement method — these are routine development events that have direct cryptographic consequences that most development teams do not monitor.

Software Bill of Materials analysis is the foundational tool for discovering transitive cryptographic dependencies. An SBOM that includes cryptographic library dependencies and their versions provides the starting point for assessing which algorithm families are in use across the application portfolio. The CBOM standard — the Cryptographic Bill of Materials — extends this concept to enumerate specific cryptographic assets, algorithms, and key material characteristics in a machine-readable format. Both standards are discussed in the tooling section of this chapter.

5.2 Discovery Methodology: A Layered Approach to Cryptographic Visibility

5.2.1 Network Scanning for Protocol and Cipher Suite Enumeration

Network-layer cryptographic discovery is the fastest way to establish a baseline view of your externally observable cryptographic posture. Tools purpose-built for TLS analysis — testssl.sh, sslyze, and commercial equivalents — can scan networks at scale and report the protocol versions, cipher suites, key exchange methods, certificate details, and signature algorithms in use on every reachable TLS endpoint. This layer of discovery is non-invasive, produces rapid results, and covers the most exposed portion of your cryptographic surface — the services that any network adversary could also enumerate.

Network scanning produces actionable findings immediately. Endpoints using TLS 1.0 or 1.1, expired certificates, certificates with small key sizes, cipher suites that prioritize export-grade algorithms, and services that still accept SSLv3 handshakes are all discoverable by passive network analysis. These findings are relevant to both near-term security hardening and post-quantum migration planning because the systems that have not kept pace with classical cryptographic best practices are typically the same systems that will require the most attention in a post-quantum migration.

The limitation of network scanning is that it sees only what the network exposes. Internal service-to-service communication over encrypted private networks, database encryption at rest, key material stored in vaults or hardware devices, and application-layer signing operations are all outside the scope of network-layer discovery. Network scanning is the starting layer of a layered methodology, not a complete solution.

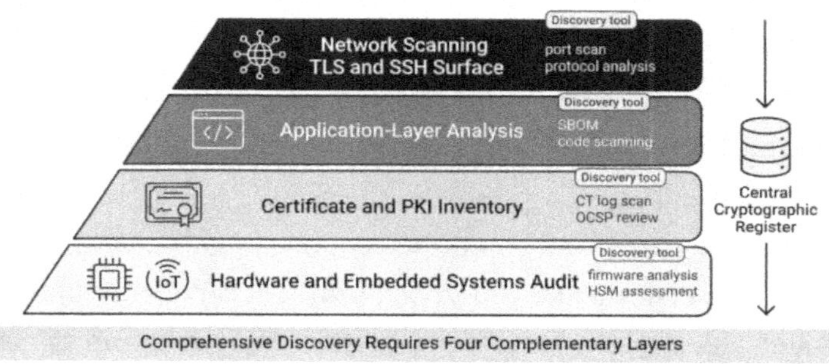

5.2.2 Application-Layer Discovery: Code Scanning, SBOM Analysis, and API Auditing

Application-layer cryptographic discovery moves inside the network perimeter to examine how applications implement cryptography in their source code, dependencies, and configuration. Static analysis tools can scan source code repositories for calls to cryptographic functions and flag the algorithms, key sizes, and implementation patterns in use. Tools like Semgrep with cryptographic rule sets, custom CodeQL queries targeting cryptographic library usage patterns, and commercial application security testing platforms with cryptographic analysis modules are the primary tools in this layer.

SBOM analysis provides a complementary view by examining the declared and transitive dependencies of each application and mapping those dependencies to cryptographic library versions. The CBOM extension layer adds cryptographic asset enumeration — specific algorithms, key types, and algorithm parameters — to the standard software bill of materials. For organizations that have not yet adopted SBOM practices, a post-quantum migration program is a compelling forcing function: the SBOM produced for migration purposes has immediate value for supply chain security, license compliance, and vulnerability management beyond its cryptographic applications.

API auditing examines the cryptographic properties of programmatic interfaces — the authentication mechanisms, token-signing methods, encryption schemes, and certificate-validation logic used by internal and external APIs. API gateways with inspection capabilities, proxy-based traffic analysis during test environments, and documentation review against implementation are the primary techniques. Organizations with large API portfolios frequently discover that different teams have made inconsistent cryptographic decisions. Some APIs use ECDSA P-256 for token signing, others use RS256 with 2048-bit keys, and others use older HMAC-SHA1 configurations that were never updated. This inconsistency is a migration planning discovery, not just a security finding.

5.2.3 Certificate Inventory: PKI Hierarchy Mapping and Expiration Risk

Certificate inventory is often the most tractable starting point for organizations beginning cryptographic discovery because certificates are the most visible and most documented cryptographic assets in most enterprise environments. Certificate monitoring tools, PKI management platforms, and certificate lifecycle management systems already exist in most mature security programs. The gap is typically in coverage and completeness rather than tooling. Internal certificates, manually issued certificates, vendor-supplied certificates, and device certificates frequently fall outside the scope of the primary certificate monitoring platform.

PKI hierarchy mapping goes beyond enumerating leaf certificates to documenting the trust chains they depend on. An enterprise PKI typically includes an offline root CA, one or more intermediate CAs, and issuing CAs for different certificate classes — TLS server certificates, client authentication certificates, code signing certificates, device certificates, and email signing certificates. Each CA in the hierarchy has its own key, its own algorithm, and its own lifetime. Migrating to post-quantum signatures requires replacing the entire chain from root to leaf — replacing only the leaf certificate while leaving the signing chain in classical algorithms provides no quantum resistance for the chain of trust itself.

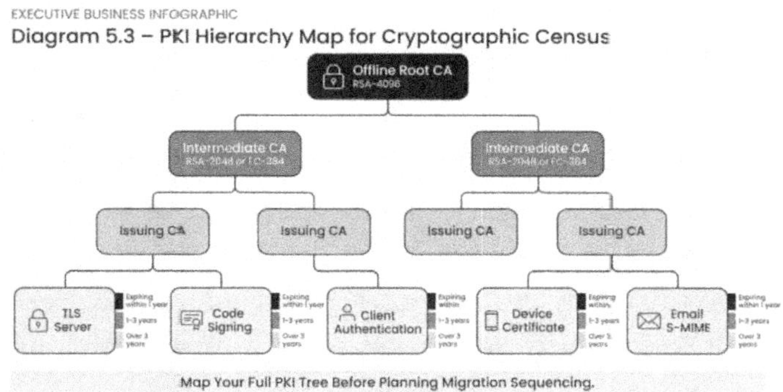

Expiration risk mapping is a byproduct of a complete certificate inventory that has immediate operational value. Certificates discovered through cryptographic census include those not in any current monitoring system — and these are frequently the most at risk of unexpected expiration. Certificates for internal services, legacy applications, device authentication, and vendor integrations often lack automated renewal processes and are renewed manually when someone notices an expiration alert. Bringing them into the census creates the foundation for extending automated lifecycle management to the full certificate population.

5.2.4 Hardware and Embedded Systems: IoT, HSMs, and Firmware-Level Cryptography

Hardware and embedded system cryptography is the hardest layer to discover and the most likely to contain surprises. Hardware devices implement cryptographic operations in firmware that is not accessible to software analysis tools. HSMs hold key material that is explicitly designed to be non-exportable. IoT devices may implement proprietary cryptographic stacks that do not conform to standard protocol behavior as observed by network scanners. Operational technology devices in industrial control environments may use vendor-specific cryptographic protocols.

Discovery in this layer requires a combination of vendor documentation review, device firmware analysis when firmware images are available, physical asset audit of critical hardware, and protocol-specific scanning for device-oriented protocols such as MQTT, AMQP, Modbus-TLS, and industrial protocol security extensions. The inventory goal for hardware is not necessarily the same granularity as for software — understanding which device models are deployed, which firmware versions are running, and what the vendor's post-quantum upgrade path is may be sufficient to assess migration feasibility, even without full cryptographic property enumeration.

Hardware security modules deserve particular attention in the census because they are typically at the center of the key management hierarchy. If your HSM does not support post-quantum algorithm generation and storage, your migration is blocked at the most critical layer, regardless of what you accomplish elsewhere in the application and network stack. HSM vendor post-quantum roadmaps should be gathered as a high-priority action in the early stages of cryptographic census work, and procurement planning should account for the possibility that HSM replacement rather than firmware upgrade is the required path.

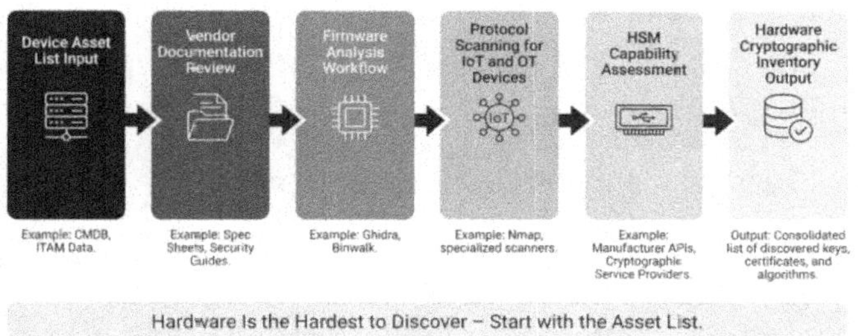

Hardware Is the Hardest to Discover – Start with the Asset List.

5.3 Vendor and Supply Chain Cryptographic Dependencies

5.3.1 Questionnaire Design for Vendor Cryptographic Posture Assessment

Your cryptographic census is incomplete if it covers only the systems your organization directly operates. Every SaaS platform, managed service, cloud infrastructure provider, and third-party integration in your ecosystem handles or transmits data that is protected — or should be protected — by cryptographic controls. The cryptographic posture of those vendors is part of your organization's effective security posture, and post-quantum migration creates explicit urgency around understanding it.

Vendor cryptographic questionnaires should be designed to elicit specific, verifiable information rather than generic assurances. The question "Does your organization use encryption for data at rest and in transit?" is not useful — every vendor will answer yes, and the answer tells you nothing about algorithm choices, key management practices, or migration readiness. Effective questions ask about specific algorithms in use, certificate authority relationships, key management platform certifications, planned post-quantum migration timelines, and FIPS 140-3 validation status for cryptographic components.

Questionnaire design should also account for the difference between what vendors know and what they are willing to disclose. Vendors may accurately report their externally visible cryptographic properties — the TLS versions and cipher suites accepted by their APIs, the certificate types used for authentication — while being unable or unwilling to describe the internal cryptographic architectures of their platforms. Building tiered questionnaires for different vendor sensitivity levels and including contractual rights to cryptographic audits in high-value vendor agreements addresses this limitation over time.

5.3.2 SaaS, Cloud, and Managed Service Provider Dependencies

Cloud infrastructure providers have generally been ahead of the enterprise in post-quantum awareness. AWS, Azure, and Google Cloud have all published post-quantum migration documentation, enabled hybrid TLS options in their respective load balancing and API gateway services, and announced timelines for post-quantum support in their key management services. Organizations that run workloads on major cloud platforms have a path to hybrid TLS protection for data in transit, available today without waiting for an internal cryptographic overhaul.

SaaS platforms present a more complex picture. A SaaS vendor that processes or stores your data uses internal cryptographic mechanisms that you cannot directly inspect or configure. Your leverage is contractual — what your service agreement requires regarding encryption standards, your right to receive security attestations and audit reports, and the migration commitments included in your vendor's published security roadmap. Cloud security assessment frameworks like CSA STAR and SOC 2 Type II reports are starting points for vendor cryptographic posture assessments, but they have historically not included post-quantum-specific questions. As regulatory guidance matures, audit frameworks will incorporate post-quantum criteria, and vendor assessments should be updated accordingly.

Managed service providers that operate your infrastructure on your behalf occupy a particularly important position in the cryptographic dependency chain because they may be responsible for key management, certificate issuance, TLS configuration, and encryption policy in your environment. MSP agreements should be reviewed for cryptographic obligations and migration accountability, and your MSP should be able to demonstrate that their managed services roadmap includes post-quantum algorithm support on a timeline that aligns with your migration plan.

Vendor Cryptographic Dependency Chain

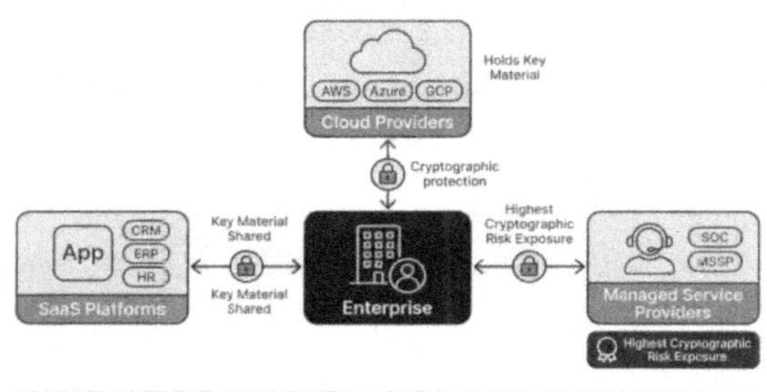

Vendor Dependencies Extend Your Cryptographic Attack Surface

5.3.3 Contractual Mechanisms for Ensuring Vendor Migration Accountability

Discovery without accountability produces inventories that generate insight but not action. The contractual dimension of vendor cryptographic posture management transforms the census from a one-time assessment into an ongoing governance mechanism. Procurement teams should work with legal and security to develop standard post-quantum migration clauses for vendor agreements, including specific algorithm requirements at defined future dates, audit rights for cryptographic posture, and service level commitments for migration milestone delivery.

Renewal leverage is one of the most effective tools for accelerating vendor migration. A vendor whose contract comes up for renewal during your migration window provides an opportunity to incorporate post-quantum requirements as a condition of renewal. This approach is particularly effective with vendors who have published post-quantum roadmaps but have not yet incorporated migration timelines into their standard service commitments. The renewal conversation converts a published roadmap into a contractual obligation, with consequences for non-delivery.

For vendors with no published post-quantum position, the contractual conversation is a risk signal regardless of outcome. A vendor that is unable to articulate a credible post-quantum migration plan for a platform that handles your sensitive data is a vendor that either lacks the technical capability or the organizational priority to perform the migration. That is information that should inform both your risk register and your vendor strategy for that platform.

5.4 From Raw Inventory to a Prioritized Cryptographic Risk Register

5.4.1 Risk Stratification: Sensitivity, Longevity, and Exposure as Scoring Dimensions

A raw cryptographic inventory is a list. A cryptographic risk register is a decision-making tool. The transformation from list to register requires applying a risk scoring framework that helps prioritize migration effort, communicate risk to stakeholders, and sequence remediation based on impact rather than convenience. The three primary scoring dimensions are data sensitivity, key longevity, and system exposure.

Data sensitivity reflects the value of the data protected by a given cryptographic asset to an adversary. Protected health information, financial credentials, intellectual property, authentication tokens for privileged systems, and personally identifiable information all carry higher sensitivity scores than publicly available content or low-risk operational data. Sensitivity scoring should align with your organization's existing data classification framework so that the cryptographic risk register integrates naturally with existing risk management processes.

Key longevity reflects how long the key material in question will be in use and how long the data it protects needs to remain confidential. A TLS certificate with a 90-day lifetime protecting public web content has low longevity risk. A root CA key protecting a hierarchy that will sign certificates for the next decade has high longevity risk. Data encrypted today that must remain confidential for twenty years has extreme longevity risk. Longevity is the dimension that makes harvest-now-decrypt-later risk concrete: long-lived data protected by classical keys is a harvest target today, not in the future.

System exposure reflects how accessible the system or data is to potential adversaries. Internet-facing systems, systems connected to partner networks, and systems accessible to contractors or vendors have higher exposure than purely internal systems accessible only to privileged users on private network segments. High-exposure systems with sensitive, long-lived data and classical cryptographic protection are the highest-priority targets in any migration program.

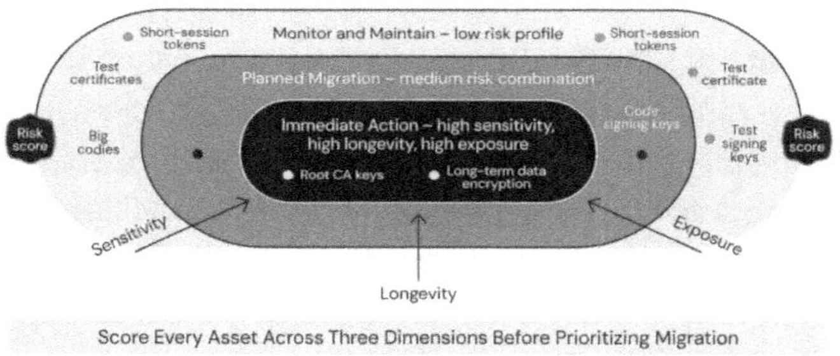

Cryptographic Risk Scoring Matrix

5.4.2 Mapping Cryptographic Assets to Business Processes and Regulatory Obligations

A risk register gains its utility from the connections it draws between cryptographic assets and the business processes and regulatory obligations that depend on them. A certificate protecting an API endpoint is not just a cryptographic asset — it is the protection mechanism for a payment processing workflow that processes several thousand transactions per hour and falls under PCI DSS scope. A signing key used for audit log integrity is not just a cryptographic asset — it is the evidentiary foundation for regulatory compliance reporting that your external auditors rely on. Making these connections explicit in the risk register transforms cryptographic risk from a technical abstraction into a business and compliance reality that executives and audit teams can understand and act on.

Process mapping for cryptographic assets requires collaboration between the security team responsible for the census and the business process owners who understand what the systems are used for. Security teams frequently know more about the cryptographic properties of a system than about the business context that makes a breach of that system consequential. Business process owners often recognize the business importance of a system but lack understanding of its cryptographic dependencies. The risk register is the artifact that bridges these two knowledge domains.

Regulatory obligation mapping adds a compliance dimension that constrains migration priority decisions. Some systems must be migrated before others, not because their risk score is highest in absolute terms, but because they are in scope for a regulatory requirement with a specific timeline. Healthcare organizations with HIPAA scope, federal contractors with CMMC requirements, and financial institutions with sector-specific security rules may find that compliance-scoped systems create fixed points in their migration sequencing regardless of relative risk scores.

5.4.3 Building a Register That Executives and Audit Teams Can Both Use

The cryptographic risk register serves two distinct audiences with different needs. Executives and program sponsors need a summary view that communicates aggregate risk exposure, migration progress, and priority decisions without requiring cryptographic literacy. Audit teams need detailed evidence of specific controls, algorithm configurations, and remediation histories that support compliance attestation. Building a register that serves both requires deliberate design rather than defaulting to the format that is most natural for the security team.

The executive view should present risk in terms of business outcomes: the number of systems with sensitive data protected only by harvest-vulnerable classical cryptography, the percentage of critical infrastructure covered by post-quantum-ready algorithms, the migration completion rate against plan, and the residual risk reduction achieved by completed migration work. These metrics tell a coherent risk story without requiring the audience to understand the difference between ML-KEM and RSA.

The audit view should present the technical specifics: algorithm names and parameter sets, certificate serial numbers and validity periods, key management system certifications, migration milestone evidence with dates and responsible parties, and change control records for completed migrations. The same underlying data supports both views if the register is structured with sufficient granularity and appropriately tagged metadata.

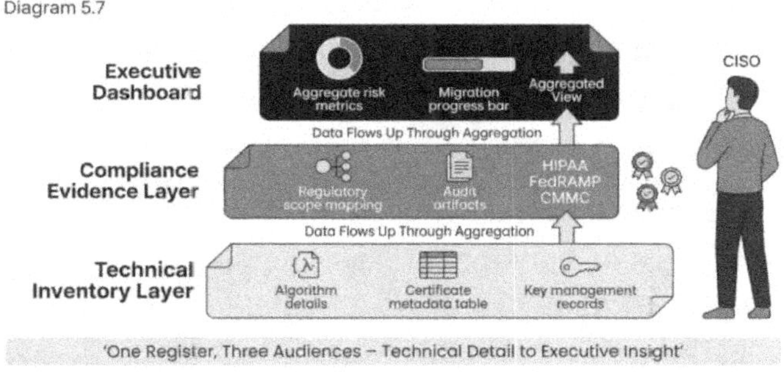

5.5 Sustaining the Inventory: Cryptographic Visibility as a Continuous Practice

5.5.1 Integrating Cryptographic Discovery into DevSecOps and Change Management

A cryptographic census, once completed and filed, produces a historical artifact that degrades in accuracy from the moment it is finished. The enterprise cryptographic environment changes continuously: new applications are deployed, certificates are renewed, dependencies are updated, infrastructure is provisioned, and vendors change their cryptographic configurations. Sustaining cryptographic visibility requires embedding discovery into the processes that drive change.

DevSecOps integration is the most effective long-term strategy for maintaining cryptographic visibility at the application layer. Build pipelines that include cryptographic property scanning as a stage alongside dependency scanning, SAST, and DAST, that produce SBOM and CBOM outputs that are updated with every build. Security gate policies that flag new or changed cryptographic library dependencies for review prevent algorithm regression—the introduction of deprecated or weak algorithms through routine dependency updates. These controls require initial engineering investment but produce continuous visibility with no ongoing manual effort once established.

Change management integration addresses the infrastructure and operational layer. Configuration management change tickets that affect TLS configuration, certificate deployments, key management systems, or network encryption should trigger cryptographic register updates as a workflow step. The goal is to make the cryptographic register a living document that reflects the current state of the environment rather than a snapshot that requires periodic re-discovery expeditions.

Diagram 5.8 – Continuous Cryptographic Discovery Integration

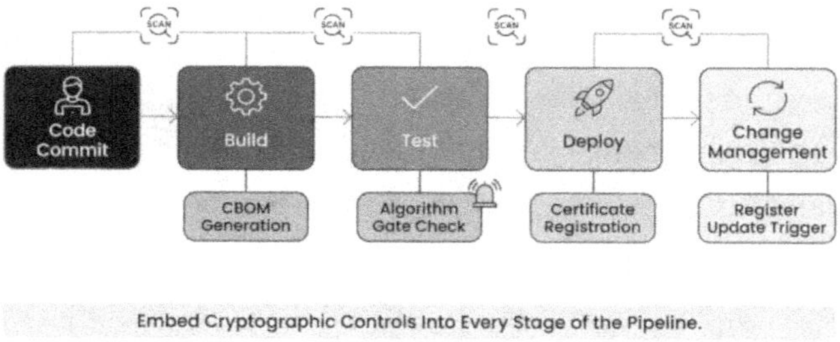

Embed Cryptographic Controls Into Every Stage of the Pipeline.

5.5.2 Tooling Landscape: CBOM, Cryptographic SBOM, and Emerging Standards

The tooling landscape for cryptographic asset discovery is maturing rapidly in response to the post-quantum migration imperative. Three categories of tools are most relevant: network protocol analyzers with cryptographic enumeration capabilities, static analysis platforms with cryptographic detection rule sets, and cryptographic bill of materials standards and associated tooling.

The Cryptographic Bill of Materials standard, developed through the CycloneDX project, extends the established SBOM format to include cryptographic assets as first-class inventory items. A CBOM entry for a cryptographic asset includes the algorithm, key size or parameter set, component role, protocol context, certificate metadata, and any associated vulnerability information. Tools that produce CycloneDX CBOM output can be integrated with existing SBOM toolchains, enabling organizations that have adopted SBOMs for software supply chain security to extend the same infrastructure to cryptographic asset management.

The National Security Agency's CBOM guidance, ENISA's cryptographic inventory recommendations, and emerging IETF work on machine-readable cryptographic posture reporting are all converging on common data formats that will enable tool interoperability and regulatory reporting. Enterprise teams that invest in CBOM tooling now will be positioned to feed automated compliance reporting as regulatory frameworks adopt these standards. The pilot-to-production pathway for cryptographic asset management tooling is shorter than for many security investments because the core technology — SBOM generation, static analysis, network scanning — is already mature; the cryptographic specialization layer is the incremental addition.

5.5.3 **Governance Structures for Keeping the Register Current**

Tools and processes sustain cryptographic visibility only within a governance structure that assigns ownership, establishes update cadences, and creates accountability for register accuracy. Without governance, even well-designed technical discovery programs produce registers that drift into inaccuracy as competing priorities absorb team attention. The governance structure for cryptographic asset management should be proportionate to the complexity of the environment and the regulatory stakes involved.

At minimum, a governance structure for cryptographic visibility requires a designated owner for the cryptographic risk register, defined responsibilities for specific asset categories among the teams that operate those assets, a quarterly review cycle that validates major register segments against current system configurations, and an incident response trigger that activates register review when algorithm-related security events occur. In larger organizations, a cryptographic steering function — whether a dedicated team or a committee drawn from security, architecture, and operations — provides the coordination layer that keeps the register aligned across organizational boundaries.

The governance structure should also define the escalation path for register gaps and anomalies. When a new system is discovered to be using a deprecated algorithm, who is notified? What is the remediation timeline? Who approves exceptions when technical constraints prevent immediate remediation? Answering these questions before they arise turns the cryptographic register from a passive inventory into an active risk-management instrument. That transformation is what makes the census investment worthwhile in operational terms.

5.6 Manager's Checklist: Running the Cryptographic Census

- Confirm that your CMDB and existing asset inventory do not cover cryptographic properties, and commission a separate cryptographic discovery effort with dedicated scope and resources.

- Schedule a network-layer TLS and SSH scan of your internet-facing systems within the first sixty days of the program to establish a baseline cryptographic posture view for externally observable services.

- Engage your development organization to identify the cryptographic library dependencies in the five highest-priority applications and produce SBOM or CBOM output for each.

- Audit your PKI hierarchy to map every CA in the trust chain, the algorithm and key size at each level, the expiration dates, and whether automated renewal processes cover each CA.

- Issue a cryptographic posture questionnaire to your top twenty vendors by data sensitivity within the first quarter of the program.
- Build a risk scoring rubric for the register that uses sensitivity, longevity, and exposure as the three primary dimensions and test it against a representative sample of discovered assets.
- Map the ten highest-risk cryptographic assets to the specific business processes and regulatory obligations they protect, and present this mapping to your CISO as the foundation for executive risk communication.
- Identify the teams and workflows responsible for the highest-priority assets and begin planning integration of cryptographic discovery triggers into their change management processes.
- Evaluate whether your CI/CD pipeline can accommodate CBOM generation as a build stage and commission a proof-of-concept in at least one development environment.
- Assign a named owner for the cryptographic risk register and define the quarterly review process, escalation path, and anomaly response procedure before the initial census is complete.

5.7 **The Census Is the Foundation**

You cannot migrate what you cannot see. The cryptographic census is not a project that runs parallel to the post-quantum migration program — it is the program's most critical prerequisite. The organizations that complete the transition with the least disruption will be those that built a comprehensive, accurate, and continuously maintained picture of their cryptographic assets before beginning systematic migration work.

The discovery methodology described in this chapter is layered by design: network scanning provides the fastest initial baseline, application analysis adds the depth that scanners cannot reach, certificate inventory closes the most common compliance gap, and hardware audit surfaces the dependencies that are hardest to migrate but impossible to ignore. No layer is optional in a serious enterprise cryptographic program, and your specific risk profile and resource constraints should determine the execution order.

The cryptographic risk register that emerges from this work is the artifact that connects the technical reality of your cryptographic environment to the business and regulatory context that should drive your migration priorities. It is the document your CISO will cite in executive briefings, the evidence your auditors will request, and the foundation for every migration decision in subsequent chapters. Invest in building it correctly, and every subsequent phase of the program becomes faster, cheaper, and more defensible.

6 Running Two Systems at Once: Hybrid Cryptography as the Bridge Across the Transition

Your network engineering team has finished configuring the first set of load balancers with ML-KEM support in a test environment. The test succeeds when the client is on a current browser version and is using hybrid TLS. It fails silently when a legacy monitoring agent tries to establish a connection using its hardcoded TLS client library. The agent predates your organization's migration planning by four years. Its vendor has announced that a post-quantum update will be available "in a future release" with no specific timeline. That agent monitors eleven production services that cannot be left unmonitored during a migration that will take at least eighteen months to complete.

This scenario captures the essential challenge of the post-quantum transition period: you cannot flip a switch and turn off classical cryptography across an enterprise of any meaningful scale. Your environment contains clients, servers, devices, libraries, and protocols at different stages of readiness, and many of those components are outside your direct control. Hybrid cryptography is the architectural response to this reality. By running a classical algorithm and a post-quantum algorithm simultaneously — combining their outputs so that the security of the result depends on the security of both — you can begin deploying quantum-resistant protection for the connections and data that matter most without cutting off the clients and systems that are not yet ready to make the full transition.

Hybrid cryptography matters because it resolves the binary that would otherwise make the transition impossible to manage safely. Without a hybrid, you face a choice: keep classical cryptography running and accept that you are accumulating harvest-now risk for as long as the migration takes, or force immediate post-quantum migration and break the systems that are not yet ready. Neither option is operationally acceptable. Hybrid provides a third path: deploy quantum-resistant protection for new connections and high-priority data while maintaining backward compatibility for systems that cannot yet be updated.

The workflow-level impact of hybrid is that it allows migration to proceed system by system, service by service, without requiring coordinated simultaneous cutover across the full environment. TLS-terminated services that support hybrid key exchange groups benefit from post-quantum forward secrecy on every connection from compliant clients today, while continuing to negotiate classical key exchange with non-compliant clients. The data is protected as quickly as the client ecosystem can support it, rather than waiting for the slowest component in the system to be updated.

The risk picture for hybrid is also mission-aligned in a specific way. If a future attack somehow breaks the post-quantum component of a hybrid construction, the classical component continues to provide the security level it always has. If a quantum computer breaks the classical component, the post-quantum component continues to protect the derived keys. The security of the hybrid construction is therefore strictly stronger than the security of either component alone — a significant property during a transition period when confidence in new algorithms is still being built through accumulated deployment experience.

6.1 The Interoperability Gap and Why a Hard Cutover Is Not Realistic

6.1.1 Mixed Environments as the Dominant Condition During Multi-Year Migrations

Enterprise infrastructure exists in a permanent state of partial modernization. At any given point in time, your organization is running systems built across multiple generations of technology, procured at different times, managed by different teams, and updated on different schedules. Some systems are on the leading edge of deployment practice. Others are running on configurations that were set during their initial deployment and have not changed materially since. This heterogeneity is not a failure of management discipline — it is the natural outcome of continuous development in a complex organization with real resource constraints.

Post-quantum migration must be planned and executed within this reality. The migration will take years for organizations of any meaningful scale. During those years, the environment will contain servers that have been upgraded to support post-quantum key exchange, servers that support only classical algorithms, clients that can negotiate hybrid TLS, clients that cannot, devices that are firmware-upgradeable, devices that are not, and vendor systems whose migration timeline is entirely outside your control. The "mixed environment" is not a temporary, exceptional state that exists only at the beginning and end of migration. It is the dominant operational condition for the entire migration period.

Architectural decisions made at the beginning of the migration must account for this multi-year mixed environment as a steady state. Security controls, monitoring capabilities, key management procedures, and incident response processes all need to function correctly in an environment where some systems have been migrated, and others have not. Hybrid cryptography is the mechanism that makes security coherent across that mixed environment — it allows the protection level of each connection to reflect the capabilities of both endpoints, rather than requiring every connection to conform to a single algorithm choice.

Diagram 6.1 – Mixed Environment During Migration Transition

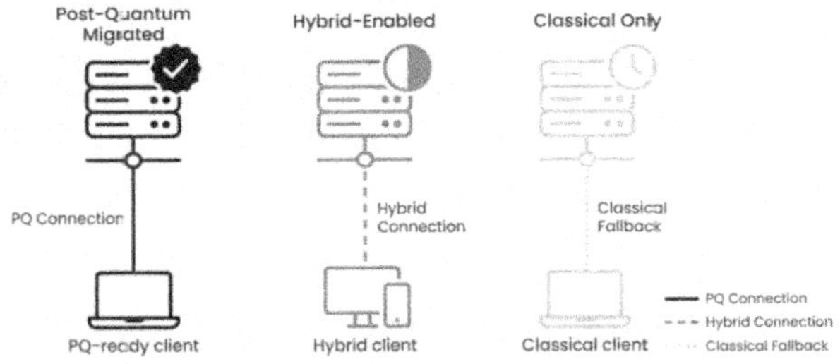

Hybrid Mode Protects Capable Connections Without Breaking Legacy Ones.

6.1.2 Backward Compatibility Requirements and Their Security Costs

Backward compatibility in cryptographic protocols is a deliberate design choice with explicit security implications. When a TLS server advertises support for both classical and hybrid key exchange groups, it can serve clients that cannot yet negotiate hybrid TLS. It also accepts the responsibility of managing the security implications of those weaker connections. A connection that negotiates classical-only ECDH, because the client cannot support ML-KEM hybrid exchange, has no quantum forward secrecy. If that client connects to a service handling sensitive, long-lived data, the connection is a harvest target.

The security cost of backward compatibility is therefore concrete and must be managed explicitly rather than assumed to be acceptable. The approach that balances security and operational continuity involves tiering compatibility policies by service sensitivity. For services handling the highest-sensitivity data, strict post-quantum or hybrid-only policies may be applied even if they break some current clients — the cost of maintaining those client connections is deemed lower than the harvest risk of continuing with classical-only protection. For services with lower sensitivity or where client ecosystem readiness is low, full classical compatibility is maintained. At the same time, hybrid support is added, and the migration of client systems is tracked separately.

Backward compatibility policies should also account for the operational risk of downgrade attacks — cases where an active adversary forces a negotiation to classical-only algorithms by interfering with the hybrid key exchange. Hybrid TLS implementations that log and alert on negotiation outcomes can detect patterns that suggest downgrade attacks, distinguishing them from legitimate compatibility fallback. Monitoring for this pattern is part of the operational discipline required by hybrid deployments.

6.1.3 **Why Hybrid Is a Bridge, Not a Permanent Destination**

Hybrid cryptography is an engineering compromise, and its costs compound over time if it is treated as a permanent architecture rather than a transition mechanism. Running two cryptographic algorithms simultaneously means larger handshake messages, higher computational load, more complex key derivation logic, larger certificates, and more complex implementations that must be maintained, tested, and validated. These costs are worth accepting during a migration period when backward compatibility is a genuine operational requirement. They are not worth accepting indefinitely.

The target architecture is a post-quantum-only environment where classical algorithms have been fully deprecated, implementation complexity has returned to a single-algorithm baseline, and the operational overhead of maintaining hybrid systems has been eliminated. Reaching that target requires driving client and server migration to the point where the remaining classical-only systems are sufficiently rare that their compatibility requirements can be addressed through specific exceptions rather than general policy. Treating hybrid as the permanent destination prevents that endpoint from being reached.

Enterprise programs should therefore define explicit hybrid deprecation criteria as part of their migration planning. At what percentage of traffic negotiating post-quantum key exchange will hybrid compatibility for a given service be retired? What is the process for communicating classical deprecation to the remaining client operators? How will exceptions be handled for clients that demonstrably cannot be updated? Answering these questions at the beginning of the hybrid deployment phase prevents the organizational inertia that keeps bridges standing long after they have served their purpose.

6.2 Hybrid Key Exchange: Combining Classical and Post-Quantum KEM

6.2.1 How Hybrid KEM Constructions Combine Key Material Safely

A hybrid key encapsulation mechanism combines the outputs of two independent KEM operations — one classical, one post-quantum — to derive a single shared secret that both parties will use for symmetric key derivation. The key property of a correctly constructed hybrid KEM is that an adversary who can break either component but not both learns nothing about the derived shared secret. If a quantum computer breaks the classical component, the post-quantum component's output provides the security. If a cryptanalytic advance breaks the post-quantum component, the classical component's output provides the security. The composition is secure under the assumption that at least one component remains secure.

The combination operation typically uses a key derivation function to combine the two component shared secrets into a single output key. The KDF must be applied in a way that ensures the security of the output depends on both inputs — simply concatenating or XOR-ing the two secrets can create constructions that are weaker than intended if one component is adversarially controlled. The IETF and standards bodies have published specific recommended combination patterns for TLS and other protocol contexts. Enterprise architects should use these standardized combination patterns rather than designing custom combination logic, as subtle errors in combination logic can undermine the security properties that make hybrid constructions valuable.

The computational overhead of a hybrid KEM is the sum of the individual component costs plus the KDF combination step. For ML-KEM-768 combined with X25519, the computational overhead per handshake is modest — benchmark data from OpenSSL and BoringSSL shows that the additional cost is typically measured in single-digit milliseconds even on constrained hardware. The bandwidth overhead of larger public keys and ciphertexts is more significant in constrained environments. A hybrid X25519+ML-KEM-768 TLS handshake transmits approximately 1,200 additional bytes compared to X25519 alone, which is negligible for most enterprise applications but material for protocols operating over low-bandwidth or high-latency links.

Hybrid KEM Key Derivation Flow

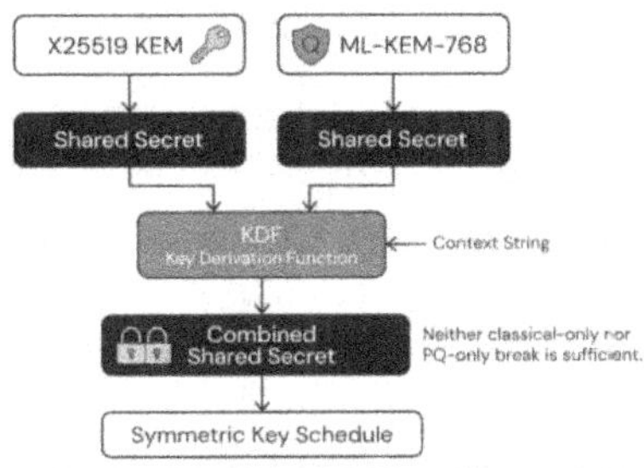

Two Algorithms, One Key – Neither Break Is Sufficient on Its Own.

6.2.2 **X25519 and ML-KEM in TLS 1.3: The Leading Hybrid Pattern**

The combination of X25519 (the elliptic-curve Diffie-Hellman function over Curve25519) with ML-KEM-768 has emerged as the leading hybrid key exchange method for TLS 1.3 deployments. This combination is defined as a named group in IETF drafts and has been implemented by Google Chrome, Mozilla Firefox, Cloudflare, AWS, and several other major TLS implementors. The wide availability of client-side support in current browser versions means that enterprises enabling this hybrid group on TLS 1.3 endpoints are immediately providing post-quantum forward secrecy for a substantial fraction of traffic from current web clients.

TLS 1.3's key exchange architecture makes hybrid integration cleaner than TLS 1.2 would have allowed. In TLS 1.3, the key exchange is the first message exchanged in a handshake, and the protocol supports negotiation of named groups that define the key exchange algorithm. Adding a hybrid group to the supported groups extension is the mechanism by which a server signals its willingness to perform hybrid key exchange. Clients who support the group will use it; clients who do not will negotiate a classical group as a fallback. This negotiation model is backward compatible by construction.

In practice, server configuration for hybrid TLS involves adding the hybrid group identifier to the TLS cipher group priority list in the server's TLS library configuration. The specifics depend on the library in use — OpenSSL, BoringSSL, and JSSE have different configuration syntax — but the conceptual operation is the same. Enterprise teams should test hybrid TLS configurations in staging environments before production deployment, particularly for services with strict latency requirements or unusual client populations. Certificate and cipher suite compatibility testing should include both current and legacy client versions.

TLS 1.3 Hybrid Key Exchange Negotiation

6.2.3 Security Analysis: What Breaks the Hybrid, and Under What Conditions

A hybrid KEM construction can be broken only if an adversary can break both component algorithms simultaneously. This is significantly harder than breaking either component alone. However, several implementation and deployment issues can undermine the security of a hybrid construction without requiring a break in the underlying mathematics.

Key reuse vulnerabilities arise when the same key material is used for multiple cryptographic operations in ways that create cross-context attacks. In the hybrid KEM context, this concern primarily concerns whether the combination function correctly domain-separates the two component secrets, preventing an adversary from using known information about one component to learn information about the combined output.

Implementations that follow the standardized combination patterns avoid this issue by construction.

Protocol downgrade attacks attempt to force a connection to negotiate a weaker key exchange group than either the server or the client would prefer. TLS 1.3 includes downgrade protection mechanisms, but these require correct implementation on both sides. Misconfigured servers that accept TLS 1.2 connections without downgrade protection may be susceptible to attacks that force negotiation away from hybrid groups. The monitoring approach for detecting downgrade patterns is discussed in the operational discipline section of this chapter.

Implementation side-channels are a risk for any cryptographic implementation. Still, they are particularly relevant to the post-quantum component of hybrid constructions because ML-KEM implementations are newer and have had less exposure to side-channel analysis than mature elliptic-curve implementations. Using well-audited, actively maintained implementations from established cryptographic libraries rather than implementing KEM operations from scratch is essential for side-channel safety. The FIPS 140-3 validation process includes side-channel resistance testing for specific implementation environments, which provides an additional validation layer for implementations in regulated contexts.

6.2.4 Implementation Patterns and Configuration Pitfalls

Several configuration pitfalls are common enough in hybrid TLS deployments that they merit explicit identification. First, load balancers and reverse proxies that terminate TLS before forwarding traffic to backend services must be updated to support hybrid key exchange, because the TLS handshake — including key exchange — occurs between the client and the termination point. If the termination point supports hybrid, but the backend connection uses only classical TLS, the end-to-end protection is determined by the weaker backend segment. Many enterprise TLS deployments use termination at multiple layers, and each termination point must be evaluated independently.

Second, certificate validation and the key exchange algorithm are independent operations in TLS. Enabling hybrid key exchange does not change the signature algorithm used to authenticate the server's certificate. The server certificate is still signed with ECDSA or RSA, and the chain of trust still uses classical algorithms. Hybrid key exchange provides quantum resistance for key derivation and forward secrecy — but not for server authentication. Full quantum resistance for authentication requires post-quantum certificates in the serving chain, which is a separate and more complex migration step.

Third, session resumption mechanisms in TLS — specifically TLS 1.3 session tickets — can interact with hybrid key exchange in ways that require careful implementation. Session tickets encode the session keys from a previous connection and allow clients to resume sessions without a full handshake. If a session ticket is decrypted using a key derived without hybrid protection and the ticket is captured by a passive adversary, resuming from that ticket provides no post-quantum security. Session ticket key rotation policies and hybrid-aware session ticket implementation are both relevant to a complete hybrid TLS security posture.

6.3 Hybrid Digital Signatures: Dual-Algorithm Signing and Verification

6.3.1 Composite Signature Schemes and Their Standardization Status

Hybrid digital signatures — composite signatures that combine classical and post-quantum signature algorithms — address the authentication dimension that hybrid key exchange does not. Where hybrid KEM protects the confidentiality of the key exchange, composite signatures protect the authenticity of certificates, documents, and signed artifacts against quantum attacks on the signing algorithm. The two mechanisms are complementary and both are needed for complete quantum resistance.

A composite signature scheme generates two signatures over the same message or data — one using the classical algorithm and one using the post-quantum algorithm — and combines them in a way that both must verify for the composite signature to be valid. The combination may be as simple as concatenation with explicit type tags, or may use a more sophisticated combination function that provides additional properties. Verifiers that support composite signatures check both components; verifiers that support only classical signatures can verify only the classical component, providing backward compatibility at the cost of quantum protection for those verifiers.

Standardization of composite signature schemes is still in progress. The IETF Lamps working group has been developing composite algorithm specifications for X.509 certificates and CMS (Cryptographic Message Syntax) signatures. These specifications define how to encode composite keys and signatures in existing certificate and signature formats without requiring changes to the fundamental certificate structure. Enterprise architects should track IETF Lamps output for composite certificate specifications as these will be the basis for vendor implementations of hybrid certificate support in PKI platforms and TLS stacks.

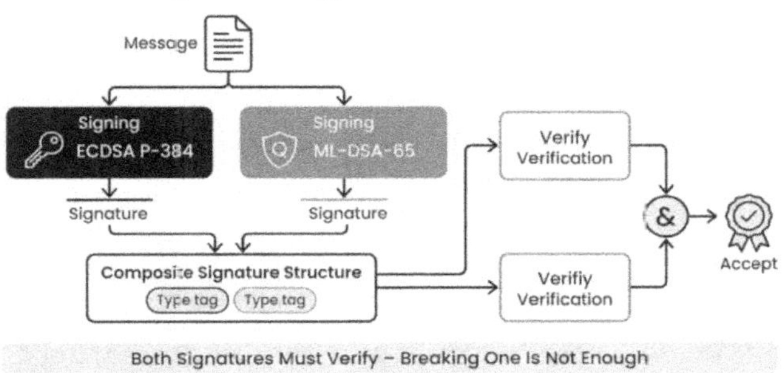

6.3.2 Certificate Chains with Hybrid Signatures: PKI Implications

Deploying hybrid signatures in a PKI context requires decisions at every level of the certificate hierarchy, not just the leaf certificate level. A web server certificate signed with a composite signature that chains to a classically signed intermediate certificate provides composite protection only for the server certificate itself. The chain of trust above that certificate still relies on classical algorithms for its integrity. A quantum adversary with the ability to forge classical signatures could produce a fraudulent intermediate certificate that chains correctly to the trusted root, regardless of whether the leaf certificate uses a composite signature.

Genuinely quantum-resistant certificate chains require post-quantum or composite signatures at every level of the PKI hierarchy. The root CA key must be a post-quantum or composite key. The intermediate CA certificates must be signed with post-quantum or composite algorithms. Only then does the entire chain derive its trust from assumptions that resist quantum attacks. This requirement significantly increases the complexity of the PKI migration and is one of the primary reasons that full PKI migration is treated as a later phase in most migration programs, typically following the hybrid TLS key exchange phase that provides earlier and lower-complexity quantum resistance for data in transit.

The practical migration sequence for PKI is typically: first, issue new intermediate CAs with post-quantum or composite keys signed by the existing classical root; second, issue new leaf certificates signed by the post-quantum intermediate; third, establish a new post-quantum root CA hierarchy signed with post-quantum keys; fourth, migrate trust anchors to the new root and deprecate the classical root. Each phase provides incremental quantum resistance while maintaining backward compatibility with clients that do not yet support post-quantum certificate validation.

PKI Hybrid Migration Sequence

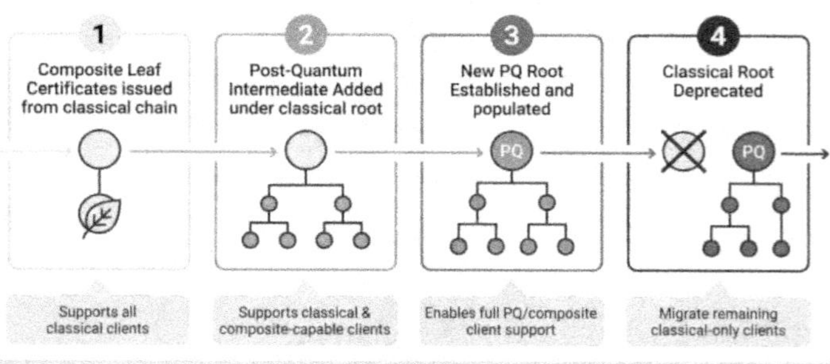

6.3.3 **Performance Trade-offs in High-Volume Signing Environments**

The performance implications of composite signatures are more significant than those of hybrid key exchange, because signing operations occur throughout the application lifecycle rather than just at connection establishment. Code signing pipelines that sign every build artifact, log integrity systems that append signatures to every log entry, transaction authentication systems that sign every financial operation, and PKI systems that issue thousands of certificates per day — all of these face amplified performance costs from composite signatures.

Benchmarking composite signature performance in your specific environment before committing to a deployment architecture is essential. The relative cost of composite signing depends on which algorithms are combined, the hardware platform in use, and whether cryptographic acceleration is available. ML-DSA signing is computationally more expensive than ECDSA on most current hardware profiles, but the gap narrows significantly on processors with vector instruction support and in environments where cryptographic hardware acceleration is available. HSMs that support ML-DSA natively — a capability that is rolling out from major vendors — provide the performance baseline for high-volume environments.

For signing environments where performance constraints genuinely preclude composite signatures, a risk-based alternative is to use ML-DSA alone — without the classical component — once confidence in the post-quantum component is sufficient. This approach accepts the loss of classical backward compatibility in exchange for full post-quantum protection with single-algorithm signing performance. The decision point for when this trade-off is appropriate depends on the maturity of your post-quantum implementation environment, the criticality of backward compatibility in your signing context, and your risk tolerance for deploying novel algorithms.

6.4 Protocol-Specific Hybrid Integration Guidance

6.4.1 TLS 1.3 Hybrid Key Exchange: Deployment and Configuration

TLS 1.3 is the most important protocol target for hybrid key exchange deployment because it protects the majority of enterprise network traffic. The deployment process for hybrid TLS 1.3 begins with verifying that your TLS library version supports the relevant hybrid key exchange groups. For OpenSSL, hybrid support is available through the OQS-Provider module, which integrates the Open Quantum Safe library's post-quantum implementations. For BoringSSL — used by Chrome, most Google services, and some commercial platforms — hybrid X25519+ML-KEM-768 support was integrated directly. For enterprise Java environments, JSSE implementations with hybrid support are available through provider extensions.

Load balancer and reverse proxy configuration for hybrid TLS requires attention to the supported groups' priority order. Placing hybrid groups at the top of the priority list ensures that clients who support hybrid will negotiate it, while clients who support only classical groups receive their best supported classical option. Monitoring tools that report per-connection negotiated groups are essential for verifying that hybrid is being negotiated at the expected rate with capable clients, and for detecting anomalies in the negotiation pattern.

Certificate rotation interacts with hybrid TLS deployment in a timing-sensitive way. The server certificate used in a TLS handshake does not need to be changed when hybrid key exchange is enabled — the certificate is used only for server authentication, and its algorithm is independent of the key exchange algorithm. This decoupling allows hybrid key exchange to be deployed immediately without waiting for the PKI migration that will eventually deliver post-quantum server certificates. Enterprises should take advantage of this decoupling to accelerate hybrid key exchange deployment before the more complex PKI migration work is complete.

6.4.2 SSH Hybrid Key Exchange: OpenSSH and Enterprise Key Management

SSH is the primary protocol for administrative access to servers, network devices, and infrastructure components, and it carries authentication credentials and session data that are high-value targets for harvest-now attacks. SSH hybrid key exchange protects the confidentiality of the session key derivation in the same way that hybrid TLS protects TLS sessions. OpenSSH has supported hybrid key exchange methods since version 9.0, released in 2022, with built-in support for sntrup761x25519-sha512 and experimental support for ML-KEM-based hybrid methods in subsequent versions.

Enterprise SSH key management introduces additional complexity for the post-quantum migration. SSH authentication typically uses public-key authentication, where users and automated systems authenticate using an SSH key pair. The authentication key pair is separate from the host key used to authenticate the server. Post-quantum SSH authentication keys — using ML-DSA or SLH-DSA signing algorithms — require key generation, distribution, and trust establishment through the same processes that manage classical SSH keys. For large enterprises with thousands of SSH key pairs distributed across servers, users, and automated systems, this is a non-trivial migration scope.

The practical recommendation for SSH is to prioritize host key algorithm migration and hybrid key exchange configuration first, since these protect all sessions regardless of client key type, and address user key migration as a subsequent step that can be phased over time. Host keys are managed centrally and updated in a defined set of server configurations. User keys are distributed across many individual credentials, making them a longer-tail migration problem that benefits from progressive enforcement through SSH configuration policies rather than forced simultaneous cutover.

Diagram 6.6 – SSH Hybrid Key Exchange Configuration

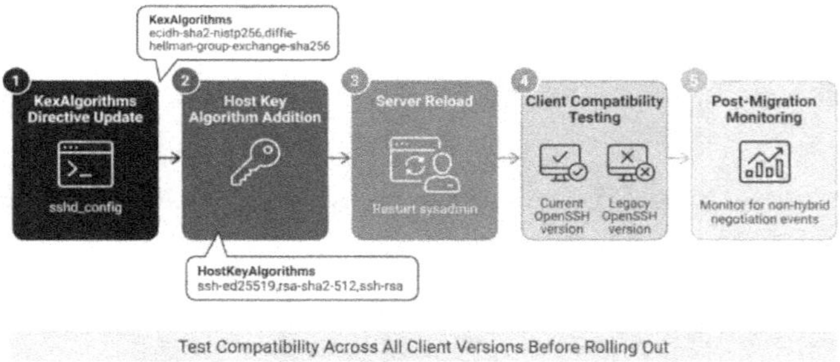

6.4.3 **Email Security: Hybrid S/MIME and PGP Migration Paths**

Email security represents one of the more challenging protocol contexts for post-quantum migration because email fundamentally differs from TLS and SSH: email messages can be encrypted and stored, then decrypted months or years later when the recipient accesses them. This long-lived nature makes email encryption a prime harvest-now target. An adversary who collects encrypted email traffic today can decrypt that traffic when quantum computers become cryptographically relevant, potentially years after the messages were sent.

S/MIME email encryption uses X.509 certificates for key encapsulation — the sender encrypts a symmetric key using the recipient's public key, and the recipient decrypts it using their private key. Migrating S/MIME to post-quantum key encapsulation requires issuing post-quantum or composite S/MIME certificates to all users who will send or receive encrypted email, updating email clients to support post-quantum key encapsulation, and managing the transition period where some users have post-quantum certificates, and others have classical certificates. The transition period requires careful handling because a classical recipient cannot decrypt a message encrypted with their post-quantum public key, and vice versa.

PGP and OpenPGP email security face similar migration challenges. The OpenPGP standard is being updated to support post-quantum cryptography through the PQC OpenPGP drafts in the IETF. Client support for post-quantum OpenPGP keys is in early implementation stages in tools such as Thunderbird, GnuPG, and OpenKeychain. Organizations with significant email encryption deployments should treat S/MIME and PGP migrations as distinct workstreams with different dependency chains and user populations, and pilot hybrid email encryption in lower-sensitivity environments before deploying to users handling the most sensitive communications.

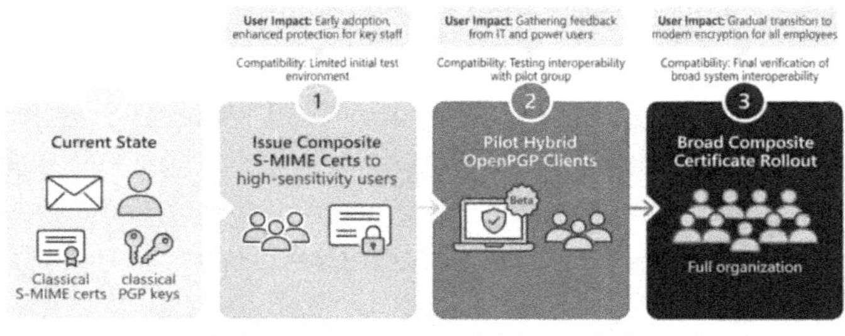

6.5 Operational Discipline for Hybrid Deployments

6.5.1 Monitoring and Alerting for Hybrid Mode Fallback Events

Hybrid TLS and SSH configurations fail silently to the observer when they negotiate classical-only connections with non-hybrid clients. This silent fallback is the correct behavior for backward compatibility. Still, it creates a monitoring challenge: you cannot distinguish between a legitimate classical fallback (a client that genuinely does not support hybrid) and a potential downgrade attack (an adversary forcing classical negotiation) without examining the negotiation logs. Operational discipline for hybrid deployments requires collecting and analyzing negotiation logs to understand the distribution of negotiated algorithms and alert on patterns that suggest manipulation.

Effective monitoring in hybrid environments tracks the per-service percentage of connections that successfully negotiate hybrid key exchange over time. A declining percentage that is not explained by known changes in the client population is an anomaly worth investigating. A service that shows 80% hybrid negotiation one week and 40% the next has likely had a configuration change, a client population change, or an active interference attempt — all of which warrant investigation. The baseline for expected hybrid negotiation rates should be established during initial deployment and updated as client populations change.

Certificate negotiation logging provides a complementary monitoring signal. TLS session logs that capture the key exchange group, cipher suite, and certificate serial number for each connection provide an audit trail to support both security monitoring and compliance reporting. For environments where all connections should eventually negotiate post-quantum key exchange, the set of connections that continues to negotiate classical groups is the priority list for follow-up — identifying those clients and driving their migration closes the remaining exposure.

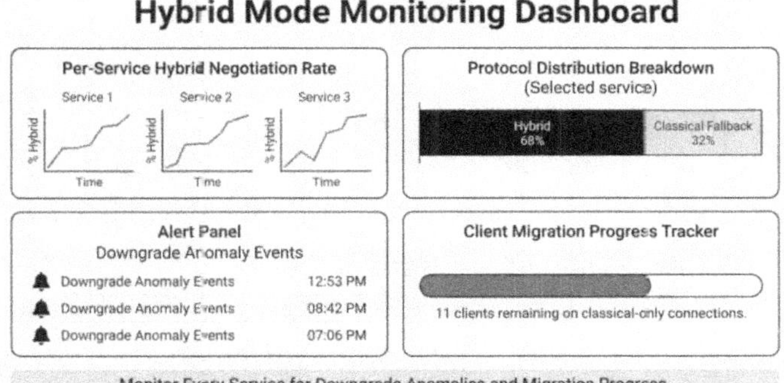

6.5.2 **Key and Certificate Lifecycle Management in Dual-Algorithm Environments**

Running two cryptographic systems simultaneously creates a key and certificate lifecycle management environment that is approximately twice as complex as managing a single-algorithm PKI. Each service may have classical certificates, post-quantum certificates, or composite certificates. Each user or device identity may have classical keys, post-quantum keys, or both. Certificate expiration monitoring must track both algorithm types. Key rotation policies must apply to both key types. Revocation processes must handle certificates from both generations.

Automation is the response to this complexity. Certificate lifecycle management platforms that support automated issuance, renewal, and revocation — ACME-compatible CAs, enterprise certificate management systems with API integration — significantly reduce the operational overhead of managing two parallel certificate populations. The investment in certificate lifecycle automation that accelerates hybrid deployment also provides lasting operational value by reducing the manual effort associated with certificate management long after the transition period ends.

Key management for dual-algorithm environments requires inventory discipline at the key level, not just the certificate level. A key management database that tracks key algorithms, key sizes, creation dates, associated certificates, authorized uses, and rotation schedules for every key in the environment supports both audit and incident response. When a vulnerability is discovered in a specific algorithm implementation, the key inventory is the tool that identifies which keys and services are affected — a capability essential for the rapid response that algorithm deprecation events demand.

6.5.3 Testing Hybrid Configurations Before Production Rollout

Hybrid cryptographic configurations introduce new failure modes that do not exist in single-algorithm environments and must be tested explicitly before production deployment. The primary test categories are: forward compatibility (do current clients successfully negotiate hybrid with the configured server?), backward compatibility (do legacy clients successfully fall back to classical negotiation?), performance under load (does the additional computational overhead of hybrid key exchange remain acceptable at expected traffic volumes?), and correctness under adversarial conditions (does the server correctly reject malformed hybrid messages?).

A staging environment that mirrors the production client population as closely as possible is essential for forward and backward compatibility testing. Synthetic client traffic using both hybrid-capable and classical-only client libraries provides a controlled test. Still, the most realistic validation comes from routing a sample of actual production traffic through the staging environment. This is particularly important for environments with diverse client populations — enterprise applications that serve mobile clients, third-party API consumers, and internal services on different update schedules will encounter the full range of TLS client behaviors only in production-representative traffic.

Performance testing for hybrid configurations should establish baseline latency and throughput metrics for the service in its current classical configuration, then measure the delta after the hybrid configuration is applied. The acceptable delta depends on the service's SLA and user experience requirements. For most enterprise services, the additional overhead of hybrid key exchange is within the tolerance of existing SLAs. For latency-critical services — real-time trading platforms, interactive video conferencing infrastructure, high-frequency API calls — the performance testing step is more consequential. It may influence the choice of the hybrid group or the timing of the hybrid deployment.

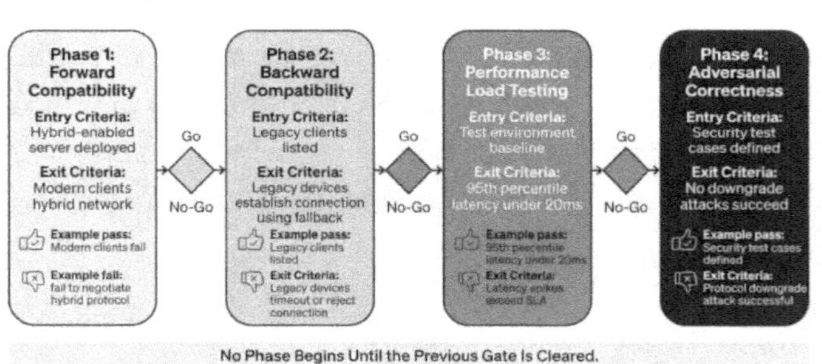

Diagram 6.9 – Hybrid Configuration Test Plan

6.6 Manager's Checklist: Deploying Hybrid Cryptography

- Verify that your TLS library versions on all internet-facing services support the X25519+ML-KEM-768 hybrid key exchange group and enable it in your staging environment within the first ninety days of the program.
- Audit your TLS termination points — load balancers, reverse proxies, API gateways — to confirm that each termination layer is updated independently for hybrid support, since the weakest termination layer determines the protection level.

- Configure TLS negotiation logging for all production services to capture the negotiated key exchange group per connection, establishing the baseline hybrid adoption rate before broad deployment.

- Deploy monitoring alerts for services where the hybrid negotiation rate drops more than ten percentage points without a known client population change, treating this as a downgrade anomaly requiring investigation.

- Update SSH host keys and KexAlgorithms configuration on all servers accessible to privileged administrators, prioritizing servers that handle the most sensitive infrastructure management operations.

- Commission a proof-of-concept for hybrid S/MIME certificates with a pilot group of users handling the highest-sensitivity email communications, documenting the client compatibility and key management lessons before broad rollout.

- Review your certificate lifecycle management platform for post-quantum and composite certificate support, and escalate to the vendor if the roadmap for this capability is not specific and credible.

- Define the hybrid deprecation criteria for each major service — at what hybrid negotiation percentage will classical fallback be turned off — and include these criteria in the migration program plan.

- Schedule a load test of the three highest-traffic services in hybrid TLS configuration before production deployment, with documented acceptable latency delta thresholds.

- Document the hybrid deployment configuration for each production service as it is deployed, including the key exchange groups configured, the fallback behavior, and the monitoring policy, creating the evidence base for compliance attestation.

6.7 The Bridge That Holds While You Build the Road

Hybrid cryptography resolves the hardest operational challenge in post-quantum migration: how to deploy meaningful quantum resistance today without breaking the systems that are not yet ready for the full transition. It does this not by finding a single algorithm that satisfies both requirements, but by running two algorithms in parallel and deriving a combined protection level that is stronger than either component alone.

The discipline required to run hybrid systems well is significant. Monitoring for fallback patterns, managing dual-algorithm certificate populations, testing configurations in representative environments before production deployment, and planning the eventual retirement of the classical components — these are operational responsibilities that do not exist in single-algorithm environments. They are costs worth accepting because the alternative — waiting for complete client ecosystem readiness before deploying any quantum resistance — accepts harvest-now risk for every day of the wait.

The bridge metaphor in this chapter's title is deliberate. Bridges are engineering solutions to the problem of crossing a gap that cannot yet be eliminated. They are designed to carry full load while permanent infrastructure is built on the other side. Hybrid cryptography serves exactly that role in the post-quantum transition. It carries your security requirements across the years-long migration gap, protecting your highest-priority systems from harvest-now risk while the full post-quantum infrastructure is built around them. Use it as intended: deploy it early, monitor it carefully, build the permanent road behind it, and plan to take it down when the road is ready.

7 Designing for Algorithmic Change: Crypto-Agility as an Enterprise Architecture Principle

7.1 **Opening Scenario**

A mid-sized financial services firm completes a two-year project to replace every instance of RSA-2048 with Elliptic Curve Cryptography. The project team celebrates. Eighteen months later, NIST finalizes its post-quantum standards, and the firm realizes it has just finished hardcoding the wrong algorithm across hundreds of applications, three internal certificate authorities, six vendor integrations, and a legacy payment gateway it cannot easily modify. The migration clock resets — but now under significantly more time pressure.

This scenario is not a failure of execution. The team did exactly what it was asked to do. The failure was architectural: no one designed the systems to accommodate the next algorithm change before the current one was complete. Crypto-agility is the discipline that prevents this pattern from repeating.

7.2 **Why Crypto-Agility Matters Now**

Cryptographic algorithms do not last forever. The shift from DES to 3DES, from MD5 to SHA-256, and from RSA-1024 to RSA-2048 each required organizations to locate every deployment of the old algorithm and replace it systematically. What made those migrations manageable —

if painful — was that each transition involved one algorithm swap at a time, and the cryptographic community typically gave organizations years between the first warning and the compliance deadline.

The post-quantum transition changes this calculus. Organizations are now facing a mandatory algorithm replacement across their entire asymmetric cryptography stack simultaneously, under a deadline driven not by standards committees but by the advancement of quantum hardware. The organizations that emerge from this transition in the strongest position will not be those that migrate fastest, but those that build systems capable of adapting to the next algorithmic change without a full rebuild.

Crypto-agility is the mission-aligned response to an environment where algorithm selection can no longer be treated as a permanent architectural decision. It is not a feature to add after the fact. It is an architectural discipline — a set of design choices, governance structures, and engineering practices that make cryptographic change a routine operational event rather than a multi-year program.

7.3 Defining Crypto-Agility Beyond the Marketing Use of the Term

The term crypto-agility appears frequently in vendor materials, policy documents, and conference presentations, often used so broadly that it loses operational meaning. For enterprise architecture, crypto-agility has a specific, actionable definition: the capacity of a system, service, or infrastructure component to have its cryptographic primitives replaced without requiring structural redesign of the application or the operational processes that depend on it.

This definition has important implications. It means crypto-agility is not a single feature but a property that must be present at every layer of the stack where cryptography operates. A TLS-terminating load balancer that supports cipher suite negotiation is agile at the protocol layer, but does nothing for the applications behind it that hardcode RSA key generation. A key management service that supports multiple algorithms is agile at the key management layer, but useless if the applications that consume keys cannot switch the algorithm they request. Agility at one layer without agility at adjacent layers creates false confidence.

Crypto-Agility Layers in Enterprise Architecture

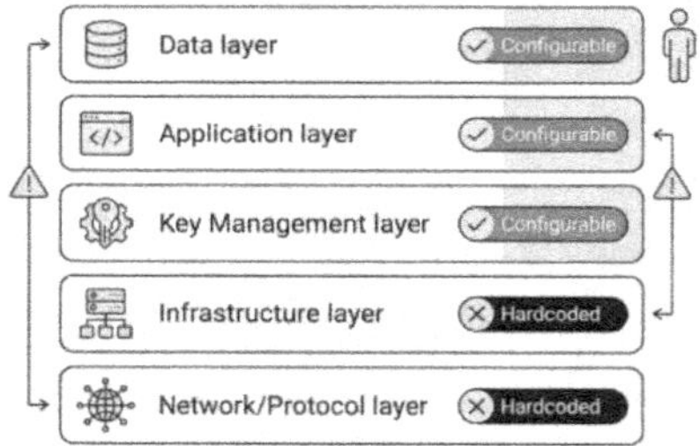

Achieving crypto-agility requires coordinated implementation across all layers of the Enterprise Architecture.

7.3.1 What Agility Means at the Protocol, Application, and Infrastructure Layers

At the protocol layer, agility means that protocol stacks support algorithm negotiation — that TLS cipher suites, SSH key exchange algorithms, and S/MIME content encryption algorithms are configurable rather than compiled-in. Most modern protocol implementations already provide this capability. Still, many enterprise deployments have constrained it through overly narrow configuration choices, legacy compatibility requirements, or misconfigured policies that permit only one cipher suite regardless of what the remote party supports.

At the application layer, agility means that cryptographic operations are not implemented in-line in business logic. When a developer needs to generate a digital signature, the application should call a well-defined cryptographic service or library interface that abstracts the underlying algorithm. When the algorithm changes, the interface contract stays the same, and only the underlying implementation changes. Applications that call OpenSSL functions directly with hard-coded algorithm identifiers, or that embed key-generation logic within business objects, are not agile. They require code changes to every affected module whenever the algorithm changes.

At the infrastructure layer, agility means that certificate authorities, Hardware Security Modules, key management services, and identity providers support algorithm diversity and can be configured to issue, store, and operate with multiple algorithm families simultaneously. Infrastructure that supports only one algorithm family at a time forces all dependent services to migrate in a single synchronized window, which is operationally equivalent to a hard cutover and eliminates the phased migration strategies that reduce risk.

7.3.2 The Spectrum from Fully Hardcoded to Fully Configurable Cryptography

Enterprise systems do not occupy a binary position of "agile" or "not agile." Most systems sit somewhere on a spectrum between fully hardcoded cryptography and fully configurable cryptography, and most enterprises have systems distributed across that spectrum with no consistent pattern.

At the hardcoded extreme, algorithm selection is baked into source code, firmware, or configuration files that are never reviewed between major releases. Key lengths, cipher suites, and algorithm identifiers appear as literals in application code. Changing them requires a code change, a test cycle, and a deployment — and often requires identifying the relevant code across dozens of repositories with no centralized registry of where cryptographic operations occur.

Toward the configurable end, algorithm selection is driven by external configuration, policy enforcement points, or service calls to a centralized cryptographic management layer. The application logic is decoupled from the cryptographic implementation. When the enterprise needs to change from ML-KEM-768 to ML-KEM-1024 or from ML-DSA to SLH-DSA, the change happens at the configuration or service layer without touching application code.

The goal is not to move every system to the fully configurable extreme — some degree of hardcoding in low-level firmware or in performance-critical paths is unavoidable and acceptable. The goal is to identify where each system sits on the spectrum, understand the cost of change at each position, and make intentional architectural investments to move high-priority systems toward the configurable end before the next migration cycle forces the issue.

7.3.3 **Measuring Your Current Agility Posture: A Diagnostic Framework**

Before building crypto-agility into new systems, enterprise architects need to understand how much agility already exists in the current environment. A practical diagnostic framework measures four dimensions: algorithm configurability, change propagation velocity, dependency isolation, and operational visibility.

Algorithm configurability measures how many steps are required to change an algorithm in production: Is it a configuration change? A library update? A code change? A hardware replacement? Systems requiring fewer steps score higher. Change propagation velocity measures how quickly a change to a shared cryptographic component propagates to all dependent systems. A key management service that can push algorithm policy updates to all consumers through a policy feed scores higher than one that requires each consumer to be individually reconfigured.

Dependency isolation measures whether a cryptographic component can be changed without requiring simultaneous changes to its consumers. If changing the signing algorithm in a certificate authority requires updating every relying party simultaneously, isolation is low. Operational visibility measures whether the organization has real-time awareness of which algorithms are in use across which systems — a prerequisite for coordinating any algorithm transition.

Running this diagnostic across your cryptographic asset inventory produces a prioritized list of agility gaps: systems that are both high-risk (handling sensitive data with long confidentiality horizons) and low-agility (requiring significant effort to migrate). Those systems represent the highest-leverage targets for architectural investment before the post-quantum deadline forces a decision.

7.4 Crypto-Agile API and Service Design

For most enterprise organizations, the largest concentration of cryptographic inflexibility is not in infrastructure — it is in the applications that call cryptographic operations directly. When developers implement signing, encryption, and key generation logic inline in application code, every algorithm transition becomes a project for application development. Crypto-agile API and service design addresses this by putting a stable interface layer between business logic and cryptographic implementation.

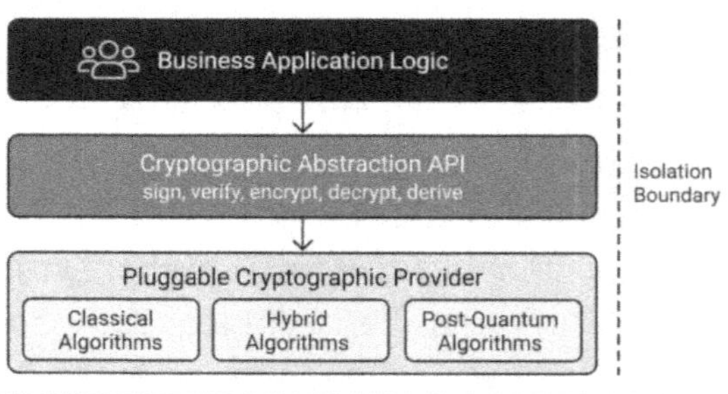

7.4.1 Algorithm Negotiation Patterns for Internal Service Communication

When two internal services communicate using encryption or digital signatures, algorithm selection should be treated as a negotiated property of the connection rather than a static configuration. The same principle that makes

TLS cipher suite negotiation valuable at the protocol layer applies to internal service-to-service communication.

A practical pattern is to define algorithm capability advertisements in service metadata, similar to how TLS ClientHello messages enumerate supported cipher suites. When Service A initiates a secure channel to Service B, Service A advertises the set of algorithms it supports and the security levels it requires. Service B selects from the intersection, preferring post-quantum algorithms when both services support them and falling back to hybrid or classical when one party does not yet support the preferred choice. This pattern allows different services to migrate to post-quantum algorithms on independent schedules without requiring coordinated cutover across all pairs.

For services that sign data consumed by downstream systems, a complementary pattern is to include algorithm metadata in the signed artifact itself. When a service produces a signed document, log entry, or data package, the signature block includes a structured identifier for the algorithm used. Downstream verifiers read the identifier and use the corresponding verification logic rather than assuming a fixed algorithm. This enables a transition period where producers migrate to new algorithms while downstream verifiers continue to support the old algorithm until all producers have migrated.

7.4.2 Abstracting Cryptographic Primitives Behind Stable Interfaces

The most durable architectural pattern for crypto-agility is the cryptographic abstraction layer: a well-defined internal API that all application code uses to access cryptographic operations. The interface exposes operations — sign, verify, encrypt, decrypt, derive key, generate keypair — without exposing algorithm identifiers to the calling code. Algorithm selection is delegated to the service implementation, which reads the algorithm policy from a centralized configuration source.

This pattern requires upfront investment in interface design, but the payoff compounds over time. Once all applications call through the abstraction layer, an enterprise can change its algorithm policy in one place — the configuration source — and the change propagates to every application that uses the service without a single line of application code changing. Testing the change requires validating the abstraction layer implementation rather than retesting every application.

The critical discipline is preventing leakage of algorithm specifics through the interface. If the interface exposes key size parameters, algorithm family flags, or format-specific encoding options, callers will eventually depend on those specifics, and the abstraction will erode. The interface should expose semantic properties — security level, performance tier, compatibility profile — and let the implementation translate those properties to algorithm selection decisions based on current policy.

7.4.3 Key Derivation and Key Management APIs That Support Algorithm Evolution

Key management APIs present a specific challenge for crypto-agility because the interface between a key management service and its consumers typically encodes algorithm-specific assumptions. Key types, key sizes, key formats, and algorithm-specific parameters are all part of the interface contract. and changing them after the fact is expensive.

An algorithm-agile key management API defines key objects by their semantic role — signing key, key-wrapping key, encryption key, authentication key — rather than by their algorithm. When an application requests a signing key, it specifies the security level and the usage context, not the algorithm. The key management service returns a key handle backed by the current algorithm for that security level and usage context. When the enterprise migrates to a new algorithm, the key management service issues new key handles backed by the new algorithm, and applications that use the handle-based API adapt automatically.

Key derivation functions present similar considerations. Key derivation chains that embed algorithm-specific parameters in derived key material can pose obstacles to migration if downstream systems depend on those parameters. Algorithm-agile key derivation APIs use abstract context identifiers rather than algorithm-specific parameters, allowing the underlying derivation algorithm to evolve without changing the interface contract.

7.4.4 Versioning Cryptographic Schemas in Persistent Data Stores

Encrypted data stored in databases, file systems, or object stores presents a unique crypto-agility challenge: the encryption algorithm becomes embedded in the stored artifact and must remain supported for as long as the artifact needs to be decrypted. Unlike connection-level encryption, where both parties negotiate in real time, stored data cannot renegotiate its algorithm after the fact.

The solution is cryptographic schema versioning. Every encrypted artifact includes a structured header that records the algorithm, key identifier, and any other parameters needed to decrypt it. When an enterprise migrates to a new algorithm, it begins encrypting new data with the new algorithm while maintaining the ability to decrypt existing data with the old algorithm. The schema version in the header determines which decryption path to use.

Migration of existing stored data can then happen asynchronously: re-encrypt on read, schedule background re-encryption for high-sensitivity data, and allow low-sensitivity data to remain under the old algorithm until its natural retention horizon. This approach eliminates the need for a synchronized migration window for stored data. It allows the enterprise to progressively reduce its exposure to the old algorithm over time rather than requiring a point-in-time cutover.

Cryptographic Schema Versioning for Stored Data

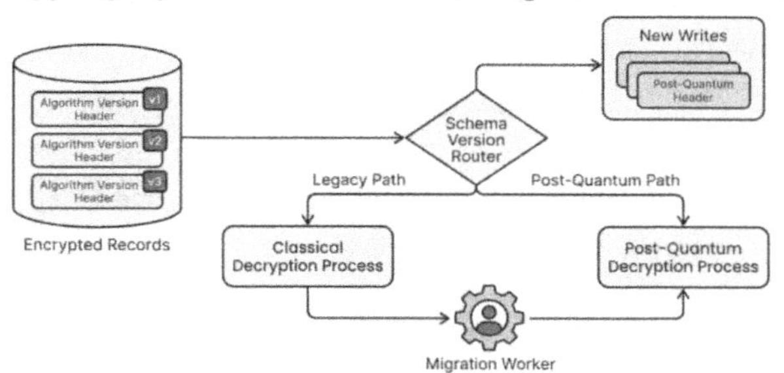

Diagram 7.2a – Cryptographic Schema Versioning for Stored Data

7.5 PKI and Certificate Infrastructure Designed for Algorithm Transition

Public Key Infrastructure is one of the most algorithm-coupled components in a typical enterprise architecture. Certificate authorities issue certificates that embed public key algorithm information; trust chains are built on algorithmic assumptions; and the relying parties that validate certificates must support the same algorithms as the certificates they validate. Making PKI infrastructure crypto-agile requires attention to certificate lifecycle automation, multi-algorithm CA hierarchies, and certificate validity period management.

Multi-Algorithm PKI Hierarchy for Transition Periods

Root CA

Classical Subordinate CA (RSA/ECDSA)

Post-Quantum Subordinate CA

Legacy Systems (not yet migrated)

Hybrid Systems (requires both trust chains)

Migrated Systems

Multi-Algorithm PKI Hierarchy for Transition Periods

7.5.1 Certificate Lifecycle Automation as a Prerequisite for Agility

Manual certificate management is incompatible with crypto-agility. When certificate operations — issuance, renewal, revocation, distribution — require human intervention at each step, the enterprise cannot rotate certificate algorithms quickly enough to respond to an algorithm deprecation event or to pilot post-quantum certificates across a subset of systems without creating operational backlogs.

Automated certificate lifecycle management, implemented via protocols such as ACME for external certificates or integrated PKI management platforms for internal certificates, is a prerequisite for certificate-level agility. Automation enables two capabilities that manual management cannot: first, the ability to change certificate templates centrally and have all subsequently issued certificates reflect the new algorithm without touching each issuing system individually; and second, the ability to shorten certificate validity periods without creating an operational burden proportional to the shortened lifetime.

Organizations that have not implemented certificate lifecycle automation should treat it as a foundational investment for the post-quantum transition — not because it directly migrates certificates to post-quantum algorithms, but because it removes the operational friction that would otherwise make algorithm migration prohibitively expensive at scale.

7.5.2 Multi-Algorithm CA Hierarchies and Their Operational Requirements

A multi-algorithm CA hierarchy maintains parallel certificate-issuance paths across different algorithm families. During the post-quantum transition, a well-designed hierarchy issues both classical and post-quantum certificates from separate subordinate CAs under a common root, allowing the enterprise to deploy post-quantum certificates incrementally while maintaining backward compatibility for systems that have not yet migrated.

The operational requirements of a multi-algorithm hierarchy are more complex than those of a single-algorithm hierarchy. The organization must maintain issuing infrastructure for each algorithm family, establish separate certificate policies for each family where requirements differ, and configure certificate distribution and validation so that relying parties receive the appropriate certificate chain for their current capabilities. Revocation infrastructure — OCSP responders, CRL distribution points — must support all active algorithm families.

The payoff is the elimination of the synchronized cutover requirement. Rather than requiring all certificate consumers to migrate simultaneously, the enterprise can migrate the PKI infrastructure first, then migrate each relying-party application at its own pace, while the multi-algorithm CA continues to serve both old and new certificate types. When the last classical certificate consumer migrates, the classical CA hierarchy can be decommissioned cleanly.

7.5.3 Short-Lived Certificates as a Migration Enabler

Certificate validity periods have historically been set at one to three years, driven by the assumption that certificate operations are expensive enough that long validity periods reduce operational burden. This assumption becomes a migration liability in an algorithm transition: long-lived certificates issued with the old algorithm remain in circulation long after the enterprise wants to complete its migration, and revoking and reissuing them all at once creates a large synchronized migration event.

Short-lived certificates — with validity periods measured in days or weeks rather than years — address this by ensuring that the certificate population naturally turns over more quickly. When the enterprise changes its certificate template to use a post-quantum algorithm, new certificates immediately use the new algorithm, and the old certificates expire naturally within the short validity window without requiring explicit revocation and reissuance.

Short validity periods require fully automated certificate renewal; manual processes cannot keep pace. But the combination of automated certificate lifecycle management and short validity periods creates a system where algorithm changes propagate through the certificate population on a timescale measured in days rather than months, which is the operational definition of certificate-level crypto-agility.

7.6 Hardware Security Modules and Cryptographic Acceleration

Hardware Security Modules represent a specific category of crypto-agility challenge because they combine algorithm dependency with hardware lifecycle constraints. An HSM that does not support a post-quantum algorithm cannot be upgraded through software alone — it may require firmware updates that the vendor has not yet released, or it may require hardware replacement if the cryptographic accelerator it uses cannot support the computational profile of post-quantum algorithms.

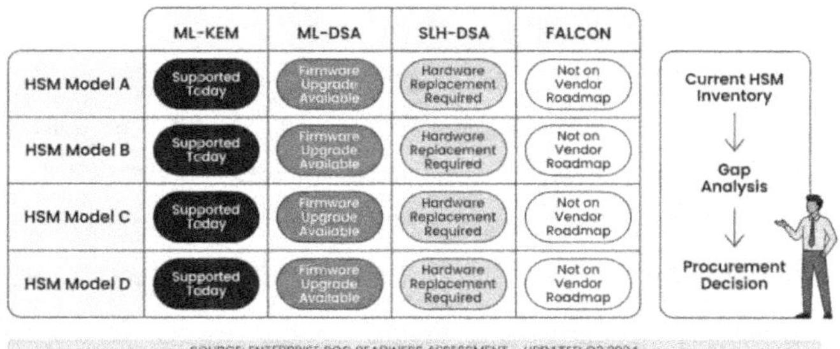

7.6.1 Evaluating HSM Firmware Upgrade Paths for Post-Quantum Algorithm Support

The first question to ask when assessing HSM crypto-agility is whether the HSM vendor has published a post-quantum algorithm support roadmap and whether that roadmap includes a firmware upgrade path for deployed hardware or requires hardware replacement. This question should be directed to vendors under a support contract and documented with a response timeline commitment, not treated as a research exercise to be revisited later.

When evaluating firmware upgrade paths, the enterprise should examine not just whether the upgrade is theoretically available but whether it is operationally feasible. HSMs in high-availability configurations may require a coordinated upgrade procedure that creates maintenance windows. HSMs integrated with hardware-validated systems — such as those underpinning FIPS 140-3-validated products — may require revalidation after a firmware change, creating a gap period during which the HSM operates outside its validated configuration.

Organizations should also examine what happens to existing keys when an HSM receives a post-quantum firmware upgrade. Keys protected under older algorithm-specific mechanisms may need to be re-wrapped under the new algorithm, or may remain in their original format indefinitely if the HSM supports parallel key stores for old and new algorithms. Understanding the key migration path is as important as understanding the firmware upgrade path.

7.6.2 Cloud KMS and Post-Quantum Readiness: Provider Assessment

Cloud-hosted key management services present a different agility profile than on-premises HSMs. The cloud provider controls the underlying hardware and can upgrade the service's cryptographic capabilities without requiring enterprise involvement in hardware lifecycle management. This makes cloud KMS architecturally more agile for algorithm upgrades — but only if the provider is actually tracking post-quantum standards and publishing a credible roadmap.

Assessing cloud KMS post-quantum readiness requires examining three things: whether the provider has publicly committed to supporting NIST post-quantum standards, what the expected timeline is for that support to reach production (not just preview or experimental), and how the provider handles the transition of existing keys and key policies when new algorithm families become available. Providers that have published specific timelines and that are already accepting test traffic on hybrid key exchange endpoints score higher on readiness than those whose only commitment is a vague roadmap item.

For enterprises using cloud KMS for workloads that handle data with long confidentiality horizons, the harvest-now-decrypt-later threat means that waiting for the cloud provider to roll out post-quantum key management before taking action is not a viable strategy. Those organizations need either to accelerate the use of post-quantum algorithms at the application layer before cloud KMS support is available, or to accept and document the residual risk of that gap.

7.6.3 **TPMs, Secure Enclaves, and Embedded Cryptographic Agility Constraints**

Trusted Platform Modules and secure enclaves in modern processors present some of the most constrained crypto-agility scenarios in the enterprise. TPMs and enclaves are designed with a specific set of supported algorithms compiled into firmware, and many of the post-quantum algorithms that NIST has standardized have computational and key-size characteristics that do not fit neatly into the fixed-size key storage and operation models that TPM specifications were designed around.

The practical implication is that TPM-backed operations — device attestation, platform key storage, Measured Boot, disk encryption — may not be directly migrable to post-quantum algorithms on current hardware. Organizations that depend on TPM-backed assurances for device trust or data protection need to understand what the hardware vendor's roadmap looks like for post-quantum TPM support, what the replacement lifecycle for endpoint hardware looks like, and whether there are software-layer mitigations that maintain the security intent of TPM-backed operations during the period before post-quantum TPM hardware is broadly available.

Secure enclaves in mobile and embedded devices face similar constraints. The cryptographic capabilities of an enclave are determined at chip design time, and post-quantum algorithm support in new silicon varies significantly across vendors and product lines. Embedded system architects working in regulated industries — medical devices, industrial control, transportation — need to factor post-quantum hardware support into their device lifecycle planning now, because the hardware procurement decisions made today determine the cryptographic agility available when the post-quantum deadline arrives.

Embedded Cryptographic Agility Constraints

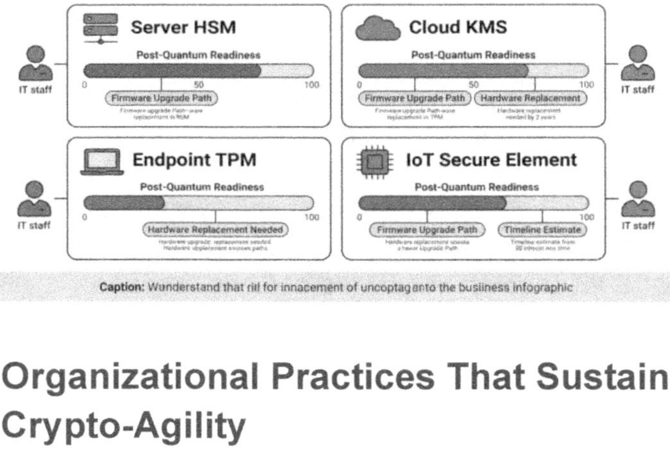

Caption: Wunderstand that ril for innacement of uncoptag anto the busiiness infographic

7.7 Organizational Practices That Sustain Crypto-Agility

Technical architecture alone does not produce crypto-agility. Systems built to be configurable become hardcoded again if organizational practices do not actively maintain the separation between algorithm policy and implementation. The technical capability to swap algorithms is only valuable if the organization has the governance mechanisms to know when to swap, the trained personnel to execute the swap, and the documentation trail to demonstrate that it was done correctly.

Crypto-Agility Governance Cycle

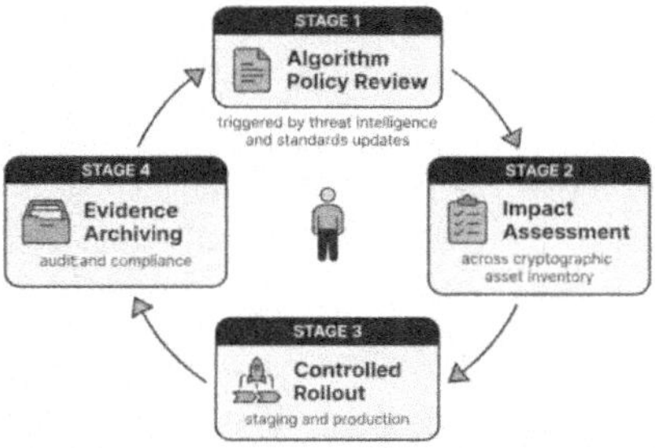

Caption: The infographic; executive business tone.

7.7.1 Cryptographic Decision Records: Documenting Algorithm Choices and Their Rationale

Every architectural decision to use a specific cryptographic algorithm — or to accept a system that hardcodes one — should be documented in a Cryptographic Decision Record. A CDR records the algorithm chosen, the use case it was selected for, the alternatives that were considered, the rationale for the selection, the expected review date, and the conditions that would trigger an earlier review. This is not bureaucratic overhead; it is the mechanism that allows future engineers and architects to understand why a decision was made, when it needs to be revisited, and what the migration path looks like when the algorithm changes.

CDRs serve a second function in organizational continuity. Algorithm decisions made by senior engineers who subsequently leave the organization are often impossible to reconstruct without documentation. When a team encounters a system using an unfamiliar algorithm or an unusual key format, a CDR explains why it was done, what the current posture is relative to the original rationale, and whether the conditions that justified the original decision still hold. Without CDRs, algorithm archaeology becomes a project in itself every time a system needs to be migrated.

CDRs should be maintained in a version-controlled repository linked to the cryptographic asset inventory. When the inventory records that a system uses a specific algorithm, the CDR linked to that algorithm decision provides the context that the inventory alone cannot capture. Together, they give the enterprise a complete picture: what is deployed, why it was deployed, and what the migration path looks like.

7.7.2 Developer Training and Secure-by-Default Cryptographic Libraries

Most cryptographic vulnerabilities in enterprise applications do not result from sophisticated attacks on well-designed systems. They result from developers making algorithm selection decisions without the context to do so well: choosing algorithm parameters by copying a Stack Overflow answer, implementing cryptographic operations in-line because a library interface was too complex, or selecting an algorithm based on performance alone without considering the security implications of the chosen key size.

Crypto-agility training for developers has a specific goal: to equip every developer who writes code that calls cryptographic operations with the knowledge to use the enterprise's approved cryptographic abstraction layer rather than an algorithm-specific implementation, and to recognize when a task requires consultation with a security specialist rather than an ad hoc implementation. This is not the same as training developers to be cryptographers. It is training developers to recognize the boundary between application logic and cryptographic implementation and to stay on their side of it.

Secure-by-default cryptographic libraries enforce this boundary at the code level. An enterprise cryptographic library that exposes only high-level semantic operations — sign, verify, encrypt, decrypt — and that automatically selects algorithms based on current enterprise policy is more agility-preserving than a library that exposes full algorithm selection to callers. When a developer calls an enterprise signing function, the library selects the algorithm. When the enterprise updates its algorithm policy, the library automatically uses the new algorithm. The developer never touches the algorithm selection code.

Secure-by-Default Cryptographic Library Architecture

Caption: Secure-by-default cryptographic library tton, and executive business infographic

7.7.3 Embedding Algorithm Review Checkpoints into Architecture Review Boards

Architecture Review Boards exist in most large enterprises to evaluate proposed system designs before they reach production. Adding a cryptographic agility checkpoint to the ARB process institutionalizes the question of algorithmic flexibility in every significant architectural decision, not just those explicitly about cryptography.

The checkpoint does not need to be complex. The ARB review template should include three questions: What cryptographic algorithms does this design use? Are those algorithms configurable without requiring code changes or infrastructure rebuild? Is the design connected to the enterprise's centralized cryptographic management and monitoring infrastructure? A design that cannot answer all three questions affirmatively either needs architectural revision before approval or needs an explicit documented exception with a remediation timeline.

Over time, this checkpoint produces a compounding effect. Every new system that passes through the ARB with a positive crypto-agility assessment does not need to be rebuilt during the next algorithm transition. Every system that fails the checkpoint and is revised before deployment is a system that does not join the inventory of hardcoded cryptography that the migration program will eventually have to address. The ARB checkpoint, in aggregate, is a mechanism for preventing the accumulation of future cryptographic technical debt.

Cryptographic Decision Record Workflow

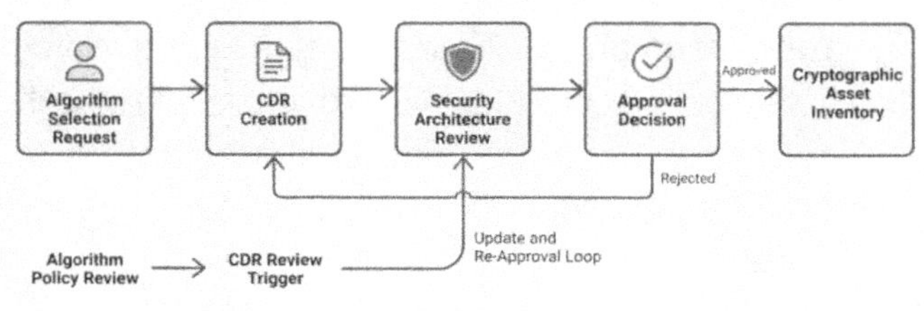

Caption: Infographic Cryptographic Decision Record Workflow with an executive business tone

7.8 Manager's Crypto-Agility Checklist

Use this checklist to assess and improve your organization's crypto-agility posture before the post-quantum migration deadline arrives:

- Run the four-dimensional agility diagnostic — algorithm configurability, change propagation velocity, dependency isolation, and operational visibility — across your top twenty most sensitive systems.
- Confirm that all new system designs reviewed by the Architecture Review Board include a crypto-agility checkpoint that addresses algorithm configurability, centralized policy connection, and monitoring integration.
- Establish or expand an enterprise cryptographic abstraction library that exposes semantic operations rather than algorithm-specific interfaces.
- Validate that your certificate lifecycle management is fully automated. If manual steps remain in the renewal or issuance process, prioritize automation before the post-quantum migration begins.

- Request post-quantum algorithm roadmaps and firmware upgrade timelines from all HSM vendors under active support contracts. Document responses and flag hardware requiring replacement.
- Assess your cloud KMS providers against the three readiness criteria: public commitment to NIST standards, production timeline, and key migration path.
- Audit your largest enterprise applications for inline algorithm selection code. Prioritize the top ten for refactoring to the enterprise cryptographic abstraction layer.
- Implement cryptographic schema versioning in all applications that store encrypted data. Verify that decryption logic supports the current algorithm and at least one prior version.
- Create a Cryptographic Decision Record process and backfill CDRs for all algorithm choices documented in the cryptographic asset inventory.
- Develop a developer training module focused specifically on identifying the boundary between application logic and cryptographic implementation, and the correct use of enterprise cryptographic libraries.
- Build a multi-algorithm subordinate CA for post-quantum certificate issuance and pilot it with a low-risk internal service before extending it to production systems.

7.9 **Key Takeaways**

Crypto-agility is not a feature to be added to a system — it is an architectural discipline that must be present at every layer of the stack where cryptographic operations occur. Protocol-layer agility without application-layer agility creates false confidence. A certificate infrastructure that supports multiple algorithms but lacks lifecycle automation cannot execute fast enough to matter in an algorithm deprecation event.

The organizational practices that sustain crypto-agility — Cryptographic Decision Records, ARB checkpoints, secure-by-default libraries — are as important as the technical architecture. Systems built with full configurability will revert to hardcoded patterns if the processes that maintain the separation between algorithm policy and implementation are not actively enforced.

The enterprises that will navigate the post-quantum transition most effectively are not those that react fastest when a threat materializes. They are those who have built the operational infrastructure — the abstraction layers, automated lifecycle management, policy-driven algorithm selection, and governance processes — that make algorithmic change a manageable, routine operational event rather than a multi-year emergency program.

8 From Inventory to Implementation: The Enterprise Post-Quantum Migration Playbook

8.1 **Opening Scenario**

The CISO of a large regional bank presents the board with a post-quantum migration roadmap. Slide four shows the list of in-scope systems: eight hundred forty-three applications, four public key infrastructure hierarchies, three cloud key management integrations, thirty-seven vendor connections requiring protocol renegotiation, and an unknown number of embedded systems in branch hardware that no one has cataloged—the slide after it shows a timeline compressed to 36 months due to regulatory pressure.

The room is silent. Not because the work seems impossible, but because no one in the room has a mental model for sequencing eight hundred forty-three parallel migration workstreams without the whole program dissolving into organizational chaos. The team needs more than a list of what to migrate. It needs a playbook: how to phase the work, how to prioritize within each phase, how to manage dependencies between systems, and how to demonstrate progress to a board that wants evidence of real completion rather than activity metrics.

This chapter is that playbook.

8.2 **Why the Sequence of Migration Decisions Determines the Outcome**

Post-quantum migration is not a single project. It is a coordinated portfolio of projects — some technical, some organizational, some vendor-driven — that must be sequenced carefully to avoid creating new vulnerabilities while closing existing ones. The wrong sequence creates dependency conflicts that stall the program, force synchronized cutovers that concentrate risk, and burn organizational capacity on low-risk systems. In contrast, high-risk systems continue to accumulate harvest-now-decrypt-later exposure.

The right sequence treats the migration as a dependency graph, not a flat checklist. Foundational infrastructure — the PKI hierarchies, the key management services, the identity and authentication layers — must migrate before the applications that depend on them, because those applications cannot switch to post-quantum algorithms until the infrastructure beneath them can issue, distribute, and validate post-quantum keys and certificates. Applications that are simple, low-risk, or already scheduled for retirement should be sequenced later, not earlier, because the opportunity cost of investing migration effort in low-risk systems is deferred progress on high-risk ones.

The playbook that follows is calibrated for large, complex organizations with heterogeneous environments. It assumes you have completed the cryptographic asset inventory and risk register described in Chapter 5, that you have at least begun the hybrid deployment preparations described in Chapter 6, and that you have the crypto-agility infrastructure decisions documented in Chapter 7 either in place or in active development.

8.3 Migration Phasing: Sequencing Decisions That Determine Outcome

The most consequential decision in a post-quantum migration program is how to phase the work. Phases are not arbitrary time slices — they are logical groupings defined by dependency relationships, risk levels, and operational complexity. A well-designed phasing structure means that each phase completes work that both reduces risk and enables the next phase. A poorly designed phasing structure means that phases overlap, dependencies are violated, and the program stalls while teams wait for infrastructure work to complete before their application migrations can proceed.

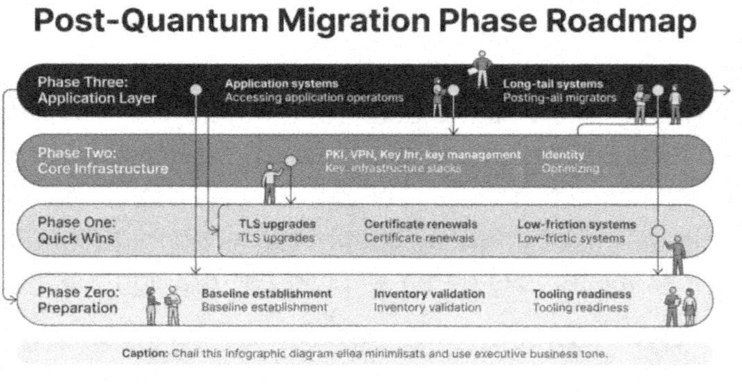

Post-Quantum Migration Phase Roadmap

Caption: Chail this infographic diagram ellea minimlisats and use executive business tone.

8.3.1 Phase Zero: Establishing the Baseline Before Touching Production

Phase Zero is the work that makes the rest of the migration possible. It is not about migrating systems — it is about ensuring the organization has the visibility, tooling, governance structures, and technical foundations needed to execute a phased migration without creating new risks.

Phase Zero deliverables include a validated cryptographic asset inventory with risk stratification, a confirmed set of post-quantum algorithm selections aligned with NIST standards and organizational policy, a hybrid deployment capability in at least the network perimeter layer, a key management service with post-quantum algorithm support in at least non-production environments, and a migration program charter with executive sponsorship, budget allocation, and cross-functional ownership defined.

The most common Phase Zero failure is treating it as a formality and compressing it to meet a program timeline pressure. Organizations that shortcut Phase Zero typically discover mid-migration that their inventory is incomplete, that their key management infrastructure cannot support the algorithm change, or that no one owns the decision authority to resolve the conflicts that arise when a production cutover affects multiple stakeholders simultaneously. Phase Zero investment pays forward across every subsequent phase.

8.3.2 Phase One: Quick Wins — Certificate Renewal, TLS Upgrades, and Low-Friction Migrations

Phase One targets migrations that combine high impact with low operational complexity: TLS cipher-suite upgrades on edge infrastructure, activation of hybrid key exchange on perimeter devices, post-quantum certificate issuance for external-facing services, and algorithm upgrades in development and staging environments. These migrations matter because they reduce harvest-now-decrypt-later exposure on the highest-visibility network paths, they generate operational experience with post-quantum configurations before tackling more complex systems, and they demonstrate program progress to executive stakeholders early in the timeline.

TLS cipher suite upgrades at the network perimeter are often the fastest Phase One win. Modern load balancers, API gateways, and CDN configurations support post-quantum hybrid key exchange groups through software or firmware updates, without requiring changes to applications. An organization can activate hybrid TLS key exchange on its external-facing TLS termination points within weeks, immediately reducing the risk that any session traffic captured after that point will be decryptable by a future quantum attacker.

Certificate upgrades for external-facing services are the second priority in Phase One. Web-facing services, external APIs, and partner integration endpoints that use publicly trusted certificates can transition to post-quantum or hybrid certificates as existing certificates expire, without requiring changes to the applications behind them. Since publicly trusted certificate authorities are already piloting post-quantum certificate issuance, this migration can often be executed entirely at the infrastructure layer without application involvement.

Phase One Quick-Win Migration Targets

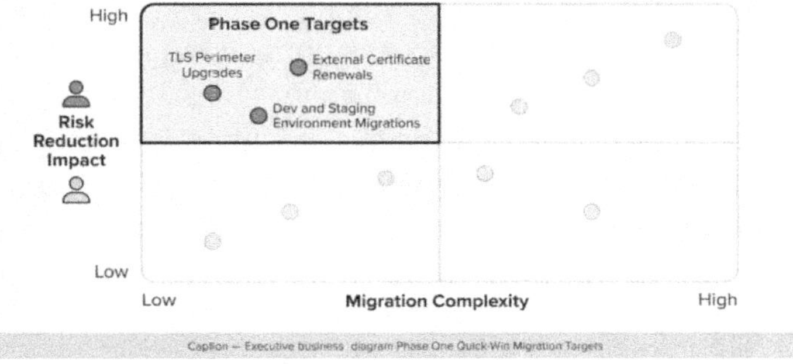

Caption — Executive business diagram Phase One Quick-Win Migration Targets

8.3.3 Phase Two: Core Infrastructure — PKI, VPN, Key Management, and Identity

Phase Two is the organizational and technical center of gravity for the migration program. It targets the foundational infrastructure that all other systems depend on: internal PKI hierarchies, VPN and remote access infrastructure, key management services, and identity and authentication platforms. These systems must migrate before the applications that depend on them can migrate, which means Phase Two work creates the enabling conditions for Phase Three.

Internal PKI migration is typically the most complex Phase Two workstream. The enterprise must stand up post-quantum CAs, update certificate templates, configure relying-party systems to trust the new CA hierarchy, and establish a transition period during which both classical and post-quantum certificates are issued and trusted. The multi-algorithm CA architecture described in Chapter 7 provides the structure for this transition. The sequencing principle is to migrate the CA infrastructure before migrating the applications, because the CA must be able to issue post-quantum certificates before any application can deploy them.

VPN and remote access infrastructure migration is often technically simpler than PKI migration, but organizationally more sensitive because it directly affects the user experience. VPN clients must support post-quantum key exchange, and the server-side configuration must be updated to prefer post-quantum cipher suites. Testing must include the full range of client platforms and operating system versions in the enterprise's environment, including legacy clients that may not yet support post-quantum algorithms and that may require hybrid fallback configurations during the transition period.

Key management service migration requires particularly careful dependency analysis. Any application that stores keys in the KMS, retrieves key material for cryptographic operations, or uses the KMS for key wrapping has a dependency on the KMS algorithm configuration. The migration sequence should update the KMS to support post-quantum algorithms first, then migrate the highest-risk key types — signing keys for code and firmware, master encryption keys for sensitive data stores — before migrating the broader population of application keys.

8.3.4 Phase Three: Application Layer and Long-Tail Systems

Phase Three addresses the application-layer migrations that cannot proceed until Phase Two infrastructure is in place, along with the long tail of lower-priority systems that do not justify the early investment of Phase One or Phase Two sequencing. This phase is typically the longest and the most organizationally complex, because it requires coordination with application development teams across the enterprise and with vendor partners whose migration timelines may not align with the enterprise's schedule.

Application-layer migration should be organized around the cryptographic dependency graph established in the asset inventory. Applications that share a PKI dependency should be grouped so that the PKI migration in Phase Two directly enables a coordinated Phase Three migration for all dependent applications. Applications that call a common cryptographic service should be migrated after the service migration, so that the algorithm change propagates to all callers through the service rather than requiring individual application code changes.

Long-tail systems — legacy applications scheduled for retirement but not yet decommissioned, IoT devices with long replacement cycles, and partner-facing integrations where migration depends on partner readiness — require a different management approach than mainstream application migrations. For these systems, the appropriate response is often a compensating control strategy: network segmentation to limit the exposure of non-migrated systems, enhanced monitoring to detect anomalous traffic patterns that might indicate active exploitation, and documented risk acceptance with an explicit remediation timeline tied to the decommission or replacement date.

8.4 Risk Prioritization Across a Heterogeneous Asset Portfolio

A migration program that treats all systems as equally urgent will waste capacity on low-risk migrations while high-risk systems continue to accumulate harvest-now-decrypt-later exposure. The risk prioritization framework that drives migration sequencing must be calibrated to the actual threat: which data, if decrypted in the future by a quantum-capable adversary, would cause the most damage to the organization, its customers, its partners, or the public interest it serves?

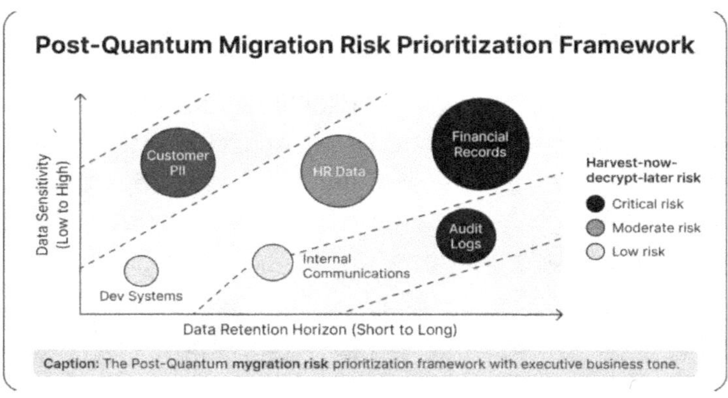

Caption: The Post-Quantum mygration risk prioritization framework with executive business tone.

8.4.1 The Harvest-Now Risk Multiplier for High-Sensitivity Data Systems

The harvest-now-decrypt-later threat changes the risk calculation for systems that handle data with long confidentiality horizons. In a conventional threat model, a system that is adequately secured against current attackers is adequately secured. In the harvest-now model, a system that is adequately secured against current classical attackers may already be generating ciphertext that is being archived by nation-state adversaries for future decryption with quantum hardware.

The harvest-now risk multiplier applies most severely to systems that handle: personal health records with decades-long sensitivity horizons; financial transaction records subject to long-term regulatory retention; intelligence and national security information with indefinite sensitivity; personally identifiable information that retains value to identity thieves long after collection; and proprietary intellectual property that competitors would benefit from even years from now.

Systems in these categories must be treated as having an effective migration deadline measured not by when quantum computers will be capable of breaking current encryption, but by when adversaries are believed to have begun archiving encrypted traffic from those systems. For systems handling national security information or long-horizon health records, the effective deadline may already have passed — adversaries may already be storing ciphertext that will eventually be decryptable. This shifts the migration priority for these systems from urgent to critical.

8.4.2 Dependency-Aware Sequencing: Migrating Foundations Before Dependents

Every enterprise has a layered dependency structure in its cryptographic architecture. Applications depend on key management services. Key management services depend on HSMs. Certificate-relying applications depend on certificate authorities. Authentication systems depend on cryptographic credential stores. A dependency-unaware migration sequence that attempts to migrate applications before the infrastructure they depend on is ready produces either partial migrations — where the application code changes but the infrastructure cannot support the new algorithm — or dangerous hybrid configurations where the application and its infrastructure use different algorithms without proper negotiation.

Mapping the dependency graph for each migration target is a Phase Zero activity that pays dividends throughout the program. A dependency map records not just direct dependencies — the PKI hierarchy that issues certificates to a specific application — but transitive dependencies: the HSM that the PKI hierarchy depends on, the key management service that manages the HSM's master keys. A migration that changes the signing algorithm of an application without first ensuring that every component in the dependency chain supports the new algorithm will fail at the component that has not yet migrated.

Dependency-aware sequencing produces a partial ordering of migration tasks: a set of constraints that say "X must migrate before Y." This ordering reduces the set of valid migration schedules and eliminates sequences that would create dependency conflicts. Within the constraints imposed by the dependency order, the migration program can optimize for other factors: concentrating effort on high-risk systems, grouping systems that share teams or vendors for efficiency, and spreading risk by avoiding simultaneous migrations of multiple critical systems.

8.4.3 When to Accept Residual Risk Versus When to Accelerate

Not every system in a large enterprise will be migrated before the theoretical deadline for quantum-capable hardware. The migration program will make triage decisions: which systems get migrated early, which systems get migrated on schedule, and which systems remain on classical cryptography longer than is ideal because their migration is complex, expensive, or blocked on vendor readiness.

The framework for making these triage decisions is explicit risk acceptance: for each system that will not meet the migration target, document the risk that has been accepted, the factors that are blocking faster migration, the compensating controls that reduce the exposure, and the conditions that would trigger an acceleration of the migration schedule. Risk acceptance is not the same as risk ignorance. It is a formal, documented decision that a specific level of residual risk is acceptable given current constraints, and that the decision will be revisited if those constraints change.

Systems that should never be candidates for risk acceptance include those that sophisticated adversaries actively target, those that handle data with indefinite sensitivity horizons, those that are required to demonstrate cryptographic compliance with specific regulatory frameworks, and those whose compromise would create cascading risk across other systems. For these systems, the appropriate response to migration blockers is escalation and resource allocation, not risk acceptance.

8.5 Vendor Readiness Assessment and Procurement Leverage

Enterprise post-quantum migration cannot proceed faster than the vendors on which the enterprise depends. Most large organizations have dozens or hundreds of software vendors, SaaS providers, managed service providers, and technology partners whose cryptographic capabilities are outside the enterprise's direct control. Managing these dependencies requires a structured approach to vendor readiness assessment, targeted use of procurement leverage, and contingency planning for vendors who fall behind schedule.

	Published Roadmap	NIST Algorithm Support	Testing Availability	Production Timeline	Migration Assistance
HSM Vendors	Ready (score 3)	In Progress (score 2)	Not Started (score 1)	In Progress (score 2)	Not Started (score 1)
PKI Vendors	Ready (score 3)	In Progress (score 2)	Not Started (score 1)	Not Started (score 1)	Not Started (score 1)
VPN Vendors	Ready (score 3)	Ready (score 3)	Not Started (score 1)	In Progress (score 2)	Not Started (score 1)
SaaS Providers	Ready (score 3)	In Progress (score 2)	Not Started (score 1)	Not Started (score 1)	Not Started (score 1)
Cloud KMS Providers	Ready (score 3)	In Progress (score 2)	Not Started (score 1)	In Progress (score 2)	Not Started (score 1)
Dev Tooling Vendors	Ready (score 3)	Ready (score 3)	Not Started (score 1)	In Progress (score 2)	Not Started (score 1)

Legend: ● Ready (score 3) ● In Progress (score 2) ○ Not Started (score 1)

Diagram 8.3 – Vendor Post-Quantum Readiness Scorecard: Vendor assessment for post-quantum cryptographic readiness

8.5.1 **Building a Vendor Post-Quantum Readiness Scorecard**

A vendor post-quantum readiness scorecard translates the question "is this vendor ready?" into a structured, comparable assessment that can be tracked over time and used to prioritize vendor management attention. The scorecard evaluates vendors across five dimensions: algorithm support status, published migration timeline, testing and pilot availability, migration assistance documentation, and contractual commitment to post-quantum readiness.

Algorithm support status measures whether the vendor currently supports any NIST-standardized post-quantum algorithms in any form — even in a beta, experimental, or limited-availability configuration. A vendor with no algorithm support in any form is at a different readiness level than a vendor with experimental support in a pre-production environment, which is at a different level than a vendor with production-ready support for at least one NIST algorithm.

Tracking progression through these levels gives the enterprise early warning of which vendors are likely to be ready on the enterprise's migration timeline and which are likely to create gaps.

Migration assistance documentation assesses whether the vendor has published practical guidance for customers performing post-quantum migrations on their platform, including upgrade guides, compatibility matrices, known limitations, and performance characteristics. Vendors with detailed migration guides are more likely to have actually tested the migration paths they describe than vendors with only high-level roadmap marketing. The presence of specific algorithm parameter documentation, test configuration files, and troubleshooting guidance is a stronger signal of real readiness than a product sheet.

8.5.2 Contractual Requirements and Renewal Leverage Points

Procurement and contract renewal cycles are the enterprise's primary leverage points for influencing vendor migration timelines. When a significant contract is up for renewal — or when the enterprise is evaluating new vendors for a category with post-quantum implications — the vendor's post-quantum readiness should be a formal evaluation criterion, with specific requirements and a documented timeline commitment.

Effective contractual requirements specify: the NIST algorithms that the vendor is required to support and by what date; the testing and certification requirements that the vendor's implementation must satisfy; the notification timeline the vendor must observe if post-quantum support will be delayed; and the remediation obligations if the vendor fails to meet the contracted timeline. Requirements that specify "post-quantum readiness" without algorithm specifics, certification requirements, or enforceable timelines are not contractually meaningful and will not create the vendor accountability the enterprise needs.

For vendors where the enterprise has significant spending leverage, contract renewal is an opportunity to extract migration timeline commitments that vendors might not otherwise make publicly. A vendor under pressure to retain a major contract relationship has an incentive to accelerate internal roadmaps to meet the contractual requirement. This leverage should be used strategically, focused on the vendors whose timelines are most likely to create migration-blocking gaps.

8.5.3 Managing Vendors Who Are Behind Schedule Without Creating Gaps

Despite best efforts at procurement leverage and contractual requirements, some vendors will be behind their committed timelines when the enterprise's migration program needs them to be ready. The enterprise needs a management approach for this scenario that maintains forward momentum without creating security gaps.

The first step is escalation: engaging the vendor at a senior level to understand the actual cause of the delay, whether it is technical complexity, resource allocation, or a deprioritization decision, and what the vendor needs to get back on schedule. Many vendor delays are remediable if the enterprise is willing to participate in a joint technical program, provide access to test environments, or accept a phased delivery of algorithm support rather than waiting for full production readiness.

When escalation does not resolve the timeline gap, the enterprise has three options: deploy compensating controls that reduce the risk created by the delayed migration, accept the residual risk with formal documentation and an updated timeline, or replace the vendor. Vendor replacement is a significant decision with its own risk and cost profile, but for critical infrastructure vendors whose continued delay creates unacceptable cryptographic exposure, it may be the only option that preserves the enterprise's migration timeline.

8.6 Testing, Validation, and Rollback Planning

Post-quantum algorithm implementations introduce new failure modes that classical testing approaches may not detect. Key sizes are larger, signature verification takes more computational time, and protocol handshakes may behave differently in high-latency or bandwidth-constrained environments. A testing approach calibrated only for classical algorithms will miss these failure modes and pass systems for production deployment that will fail under real-world conditions.

Post-Quantum Testing and Validation Pipeline

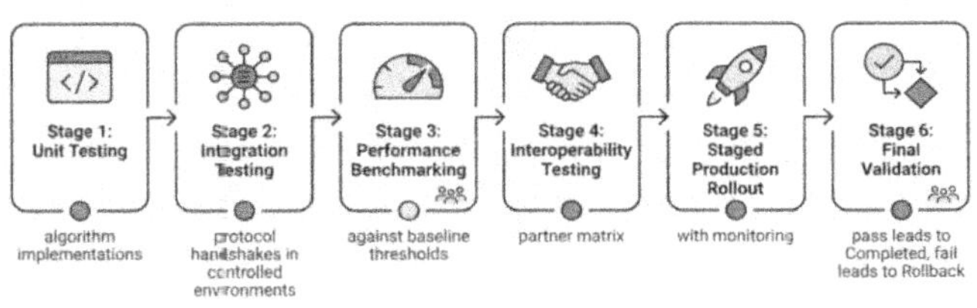

Caption: Executive business infographic diagram detailing the 'Post-Quantum Testing and Validation Pipeline'

8.6.1 **Functional Testing for Post-Quantum Algorithm Implementations**

Functional testing for post-quantum algorithm implementations must validate not just that the cryptographic operations produce correct output — key generation, encapsulation, decapsulation, signing, verification — but that the implementation handles all the failure cases that arise in real enterprise environments: malformed input data, key material in unexpected formats, algorithm parameter mismatches between sender and receiver, and fallback behavior when a post-quantum algorithm is requested but the remote party does not support it.

Test suites for post-quantum algorithms should draw on the known-answer test vectors published by NIST for each standardized algorithm. These test vectors provide unambiguous reference inputs and expected outputs that allow the enterprise to verify that its implementation of ML-KEM, ML-DSA, or SLH-DSA produces results consistent with the reference implementation. An implementation that passes known-answer tests is not guaranteed to be correct in all cases, but one that fails them is definitively incorrect.

Negative testing — deliberately providing malformed inputs, mismatched parameters, and boundary-condition values — is as important as positive testing. Post-quantum algorithms that handle edge cases incorrectly may produce incorrect output without raising an error, creating a silent failure mode that is much more dangerous than an explicit error. Test coverage should include inputs at the boundary of valid key sizes, signature lengths at and slightly beyond specified maximums, and protocol messages with truncated or extended algorithm parameter blocks.

8.6.2 **Performance Benchmarking: Establishing Baselines and Acceptable Thresholds**

Post-quantum algorithms have different performance characteristics than the classical algorithms they replace. ML-KEM has larger key sizes and ciphertext sizes than ECDH but competitive computational performance. ML-DSA signatures are significantly larger than ECDSA signatures. SLH-DSA signatures are much larger still. These characteristics affect network bandwidth, storage requirements, and the latency of cryptographic operations in high-throughput systems.

Performance benchmarking for the migration should establish baseline measurements for the systems being migrated under current (classical) algorithms, then measure the same systems under post-quantum or hybrid algorithms and identify where the performance impact exceeds acceptable thresholds. Thresholds should be defined before benchmarking begins, not after, to prevent motivated reasoning that retroactively accepts poor performance because the measurement was inconvenient.

The most common performance surprises in post-quantum migrations involve TLS handshake latency in high-connection-rate environments, certificate chain validation times when post-quantum certificates contain larger public keys, and signature verification throughput in code signing and firmware validation workflows. Each of these should be specifically benchmarked in the context of the systems being migrated, using representative workloads rather than synthetic microbenchmarks that may not reflect production behavior.

8.6.3 **Interoperability Testing in Mixed-Migration Environments**

During the migration period, the enterprise will operate in a mixed environment, with some systems migrated to post-quantum or hybrid algorithms and others not. Every interaction between a migrated and an unmigrated system must be explicitly tested to ensure that the hybrid fallback behavior is correct and that migrating one system does not break its interactions with systems that have not yet migrated.

Interoperability testing requires a partner matrix: a structured inventory of every system-to-system interaction that crosses an algorithm boundary, paired with a test that validates the interaction under the hybrid configuration that will be in place during the migration period. This matrix expands significantly in complex environments with many peer systems and partner connections, but it cannot be compressed without accepting untested migration paths that may fail in production.

Particular attention should be paid to edge cases in hybrid negotiation: what happens when a client that has been migrated to prefer post-quantum algorithms connects to a server that has not yet been migrated; what happens when a migrated server advertises post-quantum cipher suites to a client running an older version of a TLS library that does not recognize the new cipher suite identifiers; and what happens when a certificate chain includes a hybrid certificate issued by a new subordinate CA but the root CA certificate is classical and the client has not yet updated its trust store.

8.6.4 **Rollback Strategies and the Importance of Reversibility**

Every production migration should have a documented rollback plan that can be executed within a defined time window if the migration causes unexpected failures. Post-quantum migrations are no exception, and rollback planning is more complex because the migration may affect multiple interconnected systems simultaneously — rolling back one system without rolling back its dependencies may leave the environment in an inconsistent state.

The design principle for rollback-ready deployments is reversibility: maintaining the ability to revert to classical algorithms at every layer of the stack for a defined period after each migration step. This means keeping classical cipher suite configurations available in protocol implementations even after post-quantum configurations are activated, maintaining classical certificate chains alongside post-quantum chains during the transition period, and avoiding irreversible key management operations — such as deleting classical key material — until post-quantum operations have been validated in production for a sufficient period.

Rollback time windows should be defined before deployment, not after an incident. A common framework defines a 48-hour immediate rollback window during which the migration can be reverted automatically, a 7-day controlled rollback window during which reversion requires change management approval, and a post-window status in which rollback is no longer expected but a reversion path still exists for emergency use. At the end of each window, the team explicitly confirms that rollback is no longer needed before the window is closed.

Migration Rollback Decision Framework

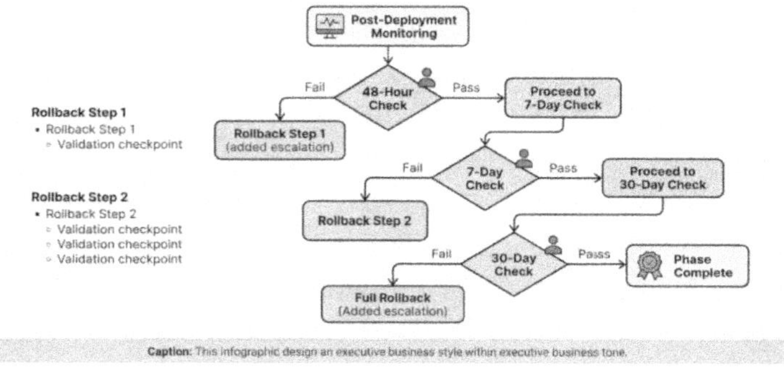

Caption: This infographic design an executive business style within executive business tone.

8.7 Change Management and Cross-Functional Coordination

A post-quantum migration program that fails at the technical level is recoverable. One that fails at the organizational level — where teams are not aligned on priorities, stakeholders do not receive timely information, and migration progress is tracked through activity metrics that do not reflect real completion — can consume years of organizational capacity without achieving meaningful risk reduction. Change management and cross-functional coordination are not soft skills peripheral to the technical program. They are the mechanisms through which technical work translates into organizational outcomes.

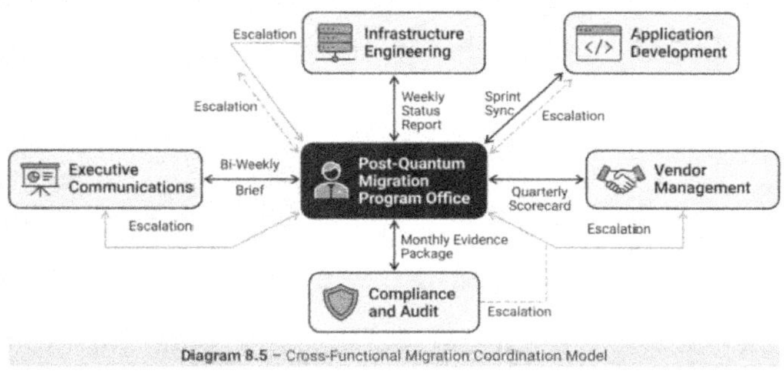

Diagram 8.5 – Cross-Functional Migration Coordination Model

8.7.1 Stakeholder Mapping for a Cryptographic Migration Program

Post-quantum migration affects more organizational stakeholders than most enterprise technology programs because cryptography is embedded in more systems than any other security control. The stakeholder map for a migration program extends beyond the security and engineering teams to include: application owners who must plan migration work in their team roadmaps; procurement and legal teams who must update vendor contracts and service agreements; compliance and audit teams who must understand the migration status relative to regulatory requirements; executive leadership who must make resource allocation decisions and accept documented residual risks; and external stakeholders including regulators, auditors, customers, and partners who may require evidence of migration progress.

Mapping these stakeholders before the program begins, rather than discovering them when their involvement becomes necessary, prevents the coordination failures that stall migration programs mid-execution. Each stakeholder group needs a defined communication channel, a defined level of information detail appropriate to their role, and a clear understanding of what decisions they are expected to make, what approvals they are expected to provide, and what they need to do to keep the program moving.

8.7.2 Communication Cadences for Leadership, Operations, and Development Teams

Different stakeholder groups need different information at different frequencies. Executive leadership needs a quarterly view of overall migration progress, risk posture, and resource adequacy. They need to understand whether the program is on track to meet its objectives, whether any escalations require their attention, and whether any decisions about budget, vendor relationships, or risk acceptance require their involvement. They do not need a detailed technical status update every week.

Operations and security teams need a weekly view of migration activity: what migrations are in progress, what is scheduled for the coming week, what issues have been identified in recent deployments, and what rollbacks or remediation actions are in progress. This cadence keeps the teams that execute migrations aligned on what is happening across the program and enables quick escalation when a deployment encounters unexpected problems.

Development teams need migration guidance that is practical and action-oriented: which systems they own are in scope for migration, what the specific technical changes required are, what the testing and validation requirements are before they can close their migration workstreams, and who to contact when they encounter blockers. Development teams typically do not need the broader program context that leadership requires, but they do need clear, specific technical requirements and a defined support channel for questions.

8.7.3 Tracking Migration Progress with Metrics That Signal Real Completion

Migration programs are prone to measuring activity rather than completion. Activity metrics — number of systems inventoried, number of migration plans written, number of configuration changes made — create a false picture of progress because they measure work done rather than risk reduced. A program that has inventoried all systems but migrated none has done necessary preparatory work but has not reduced the harvest-now-decrypt-later risk of any single system.

The metrics that signal real completion are outcome-based: the percentage of high-risk systems actively using post-quantum or hybrid algorithms in production, the percentage of TLS connections using post-quantum key exchange on external-facing services, the percentage of certificates in circulation that use post-quantum or hybrid algorithms, and the number of systems that have been verified to successfully roll back to classical algorithms within the defined rollback window.

Progress dashboards that present these outcome metrics, broken down by risk tier and by migration phase, give leadership the information they need to assess whether the program is reducing risk at the rate required by the threat timeline. A dashboard that shows eighty percent of systems with migration plans but only fifteen percent of high-risk systems with completed migrations is communicating a critical gap between planning activity and execution outcomes.

Monthly reporting to executive leadership should anchor on outcome metrics and clearly distinguish between systems where the migration is genuinely complete — post-quantum algorithms active, classical fallback removed, rollback window closed — and systems where migration is in progress but not yet complete. This discipline prevents the "paper migration" problem, where teams mark systems as migrated in the tracking tool while classical algorithm fallbacks remain active in production.

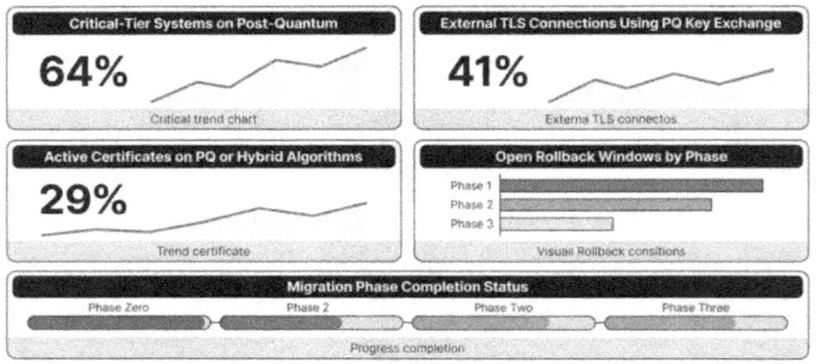

Migration Progress Outcome Metrics Dashboard

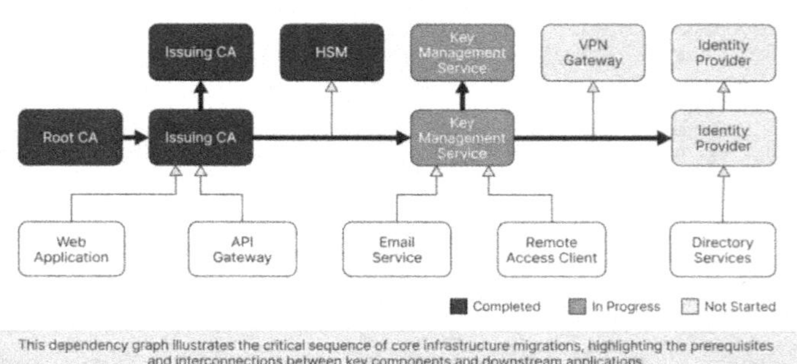

Dependency Graph for Core Infrastructure Migration

This dependency graph illustrates the critical sequence of core infrastructure migrations, highlighting the prerequisites and interconnections between key components and downstream applications.

8.8 **Manager's Migration Playbook Checklist**

Use this checklist to assess and drive your post-quantum migration program execution:

- Complete Phase Zero before any production migrations begin. Confirm that the cryptographic asset inventory is validated, the migration program charter is signed, and the key management infrastructure has post-quantum algorithm support in non-production.
- Define your Phase One target list: systems where a post-quantum or hybrid migration can be completed in under sixty days with no application code changes. Confirm these targets represent meaningful risk reduction, not just easy wins.
- Map the dependency graph for all Phase Two core infrastructure migrations. Confirm that no application-layer migration in Phase Three will be blocked by incomplete Phase Two infrastructure work.
- Assign harvest-now risk multipliers to all high-sensitivity data systems. Systems with indefinite-horizon data sensitivity should be escalated to Phase One or early Phase Two regardless of migration complexity.
- Issue formal post-quantum readiness assessments to all Tier 1 vendors. Set a response deadline and a follow-up escalation path for vendors who do not respond or who report critical gaps.
- Update contract renewal templates to include post-quantum algorithm support requirements with

specific algorithm identifiers, certification standards, and timeline commitments.

- Define performance benchmark thresholds for all high-volume cryptographic operations before beginning benchmarking. Ensure test environments use representative production workloads.
- Build the interoperability test matrix for all system pairs that cross an algorithm boundary during the migration. Test every pair explicitly before any production deployment.
- Define rollback windows and rollback execution procedures for every planned production migration. Confirm that classical algorithm configurations are maintained in all systems for the duration of the rollback window.
- Build the migration progress dashboard with outcome metrics — not activity metrics — as the primary indicators. Present outcome metrics to executive leadership at least quarterly.
- Conduct a Phase Two retrospective before beginning Phase Three. Document lessons learned from PKI, VPN, and key management migrations and apply them to the application-layer migration approach.

8.9 **Key Takeaways**

A post-quantum migration playbook is not a technology deployment plan. It is a risk management program that coordinates technical, organizational, and vendor dimensions simultaneously, in a sequence determined by dependencies and risk priorities rather than by convenience or organizational politics.

The most important execution decisions are sequencing decisions. Infrastructure before applications. High-risk systems before low-risk systems. Dependency foundations before dependent services. Compensating controls for systems that cannot migrate on schedule, rather than deferring without documentation. These sequencing disciplines are what separate a migration program that systematically reduces risk from one that generates impressive activity metrics while leaving the highest-risk systems unchanged until the last possible moment.

Migration completion requires outcome metrics, not activity metrics. A system is not migrated until post-quantum algorithms are active in production, the rollback window has been closed, and any residual classical algorithm fallbacks have been removed or documented as accepted risk. Anything short of that is work in progress, and work in progress does not reduce the harvest-now-decrypt-later risk that is accumulating in encrypted archives today.

9 Quantum Resilience as an Ongoing Practice: Governance, Workforce, and the Long Game

9.1 **Opening Scenario**

Three years after completing its post-quantum migration program, a large healthcare network discovers that a widely-used open-source library has implemented ML-DSA with a parameter derivation error that reduces the effective security level of signatures in a specific use case — one that happens to match the way the network's clinical data exchange platform signs messages. A patch is available. The organization's cryptographic governance team identifies the exposure within forty-eight hours of the advisory, activates the library replacement procedure using the update runbook built during the migration program, and completes remediation across all affected systems in eleven days without disrupting clinical operations.

For a different organization in the same situation, the same disclosure would require eight weeks to identify all affected systems, because no one maintained the cryptographic asset inventory after the migration program closed. Remediation takes another four months and requires coordination across three separate technology teams.

The difference between these outcomes is not luck, and it is not the number of resources the organizations allocated to the migration program. The difference is whether the first organization treated quantum resilience as an ongoing institutional practice or the second treated it as a one-time program. This chapter addresses what it takes to sustain the first posture over the long term.

9.2 Why Quantum Resilience Requires Ongoing Practice

The completion of a post-quantum migration program is not the end of the quantum security problem. It is the establishment of a new baseline. Cryptographic standards continue to evolve. New research has uncovered flaws in algorithms that were previously considered sound — the SIDH break described in Chapter 3 is a recent example, and it is unlikely to be the last. Threat actors continue to advance their capabilities. Regulatory requirements expand and change. Vendor implementations accumulate vulnerabilities. And the enterprise itself changes: new applications, new vendors, new data flows, and new partnerships all introduce cryptographic dependencies that were not present when the migration program concluded.

Sustaining quantum resilience requires treating cryptographic security not as a project outcome but as an institutional capability: a set of governance structures, workforce competencies, monitoring mechanisms, and response procedures that continuously maintain the organization's cryptographic posture against an evolving threat landscape. This is a more demanding institutional commitment than a one-time migration, but it is the only posture that provides durable protection.

The good news is that the organizational infrastructure built during the migration program — the cryptographic asset inventory, the governance structures, the trained personnel, the tooling — provides the foundation for ongoing practice. The investment made in building it does not depreciate immediately after the migration program closes. But it requires active maintenance to remain valuable, and organizations that let it atrophy will find themselves having to rebuild from scratch at the worst possible time.

9.3 Building a Post-Quantum Cryptographic Governance Framework

Governance is the institutional machinery that translates security intent into consistent organizational behavior. Without a governance framework, cryptographic decisions are made ad hoc, accountability is diffuse, and the organization cannot demonstrate to auditors, regulators, or board members that its cryptographic posture is being managed systematically. A post-quantum cryptographic governance framework defines who is responsible for what, what policies govern cryptographic decisions, how compliance is measured, and how cryptographic risk is communicated to leadership.

Diagram 9.1 – Post-Quantum Cryptographic Governance Framework

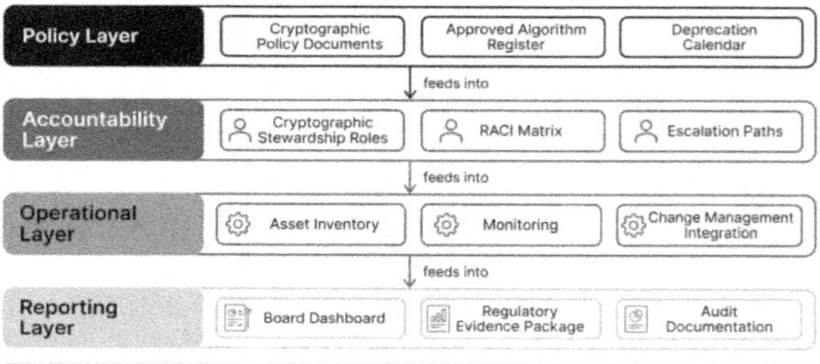

Diagram 9.1 – Post-Quantum Cryptographic Governance Framework

9.3.1 **Policy Architecture: What Documents Need to Exist and Who Owns Them**

A complete post-quantum cryptographic policy architecture comprises several interlocking documents at varying levels of specificity. At the highest level, a cryptographic security policy establishes the organization's principles for selecting cryptographic algorithms, managing keys, managing the certificate lifecycle, and retiring algorithms. This document is owned by the CISO and approved by executive leadership. It changes infrequently — typically when major standards updates occur or when significant threat intelligence warrants a policy revision.

Supporting the cryptographic security policy, a set of technical standards documents specifies the approved algorithms, key sizes, parameter sets, and protocol configurations for each use case category. These standards are owned by the security architecture team and updated on a defined review cycle—typically annually or immediately upon a relevant algorithm deprecation or standards update. The standards documents are more specific than the policy: they specify, for example, that key exchange in new TLS deployments must use ML-KEM-768 or higher, or that code signing must use ML-DSA with a security level of at least Category 3.

At the most specific level, implementation guidelines provide the practical technical details that engineering teams need to implement the standards: which library versions satisfy the standard, which configuration parameters to use, and what testing must be performed before a new cryptographic implementation is deployed to production. Implementation guidelines are owned by security engineering and updated as tooling and library ecosystems evolve.

The policy architecture is only effective if each document has a named owner, a defined review cycle, a change management process, and a distribution mechanism that ensures relevant teams are working from current versions. Documents that exist but are never updated, or that are updated but not effectively distributed, provide compliance theater rather than actual governance.

9.3.2 Roles and Accountabilities: The Cryptographic Stewardship Function

The cryptographic stewardship function is the set of organizational roles responsible for maintaining the enterprise's cryptographic posture over time. In large organizations, this may be a named team or center of excellence. In smaller organizations, it may be a set of responsibilities distributed across existing roles. The critical requirement is that the responsibilities are explicitly assigned, not assumed to be covered by general security team duties.

Core cryptographic stewardship responsibilities include: maintaining the cryptographic asset inventory; monitoring algorithmic threat intelligence and standards developments; managing the approved algorithm register and deprecation calendar; reviewing new system designs for cryptographic compliance before production deployment; coordinating vendor readiness assessments; managing the cryptographic evidence package for audit and regulatory purposes; and escalating algorithmic events — breaks, deprecations, or standards changes — to the appropriate response function.

The stewardship function requires both technical depth and organizational positioning. Technical depth is required because the team must evaluate algorithmic security claims, assess implementation vulnerabilities, and translate standards documents into actionable technical requirements. Organizational positioning requires the team to have the authority to block non-compliant deployments through the Architecture Review Board, to require vendors to meet cryptographic standards as a condition of procurement, and to escalate unresolved compliance gaps to executive leadership.

9.3.3 Integrating Quantum Resilience into Existing GRC Programs

Most large enterprises already have a Governance, Risk, and Compliance program that manages information security risk across the organization. Rather than building a parallel governance structure for quantum resilience, the more sustainable approach is to integrate quantum resilience into the existing GRC program as a risk domain with its own control framework, risk register, and monitoring mechanisms.

Integration means adding cryptographic algorithm compliance as a risk category to the enterprise risk register, with specific risk metrics tracked and reported alongside other information security risks. It means adding algorithm compliance controls to the enterprise control framework and including them in the scope of internal audits and external assessments. It means including cryptographic posture assessments in the vendor risk management program so that third-party cryptographic risks are managed with the same rigor as other third-party risks.

The advantage of GRC integration over a parallel governance structure is institutional durability. GRC programs have defined budgets, established audit relationships, and board-level visibility. Integrating quantum resilience into an existing program with those characteristics is more likely to sustain executive attention and resource allocation than a stand-alone quantum security program that competes for attention with the many other priorities on the security agenda.

9.3.4 Board and Executive Reporting: Making Cryptographic Risk Visible at the Top

Boards and executive teams have limited bandwidth for technical detail, but they have a clear need to understand whether the organization's cryptographic posture poses material risk. The governance framework must include a reporting mechanism that translates cryptographic security status into the language of business risk: what data is at risk, from what kind of adversary, with what probability and consequence, and what the organization is doing to reduce that risk on what timeline.

Effective executive cryptographic risk reporting covers three things: current posture — the percentage of high-risk systems operating on post-quantum or hybrid algorithms relative to target; emerging threats — any significant algorithm breaks, regulatory developments, or threat intelligence updates that change the risk calculus; and resource and decision needs — any escalations requiring executive attention, including vendors not meeting contractual commitments, budget gaps that would delay migration targets, or risk acceptance decisions that require executive sign-off.

Quarterly reporting to the board-level risk committee provides the cadence for sustained executive attention without creating the reporting fatigue that comes from monthly technical status updates. The quarterly report should be brief — two to three pages — and should focus on outcomes and trends rather than activity. A board that understands that 73% of critical-tier systems are on post-quantum algorithms, up from 58% last quarter, and that the remaining 27% are on track for completion within six months has the information needed to provide meaningful oversight without requiring technical expertise.

9.4 Audit Readiness and Evidence of Sustained Compliance

Regulatory auditors and internal audit teams will increasingly expect organizations to demonstrate not just that they completed a post-quantum migration but that they have sustained their cryptographic posture over time, that new systems deployed after the migration meet the same algorithm standards as migrated systems, and that they have the monitoring and response capabilities to detect and address algorithm violations when they occur.

Cryptographic Compliance Evidence Architecture

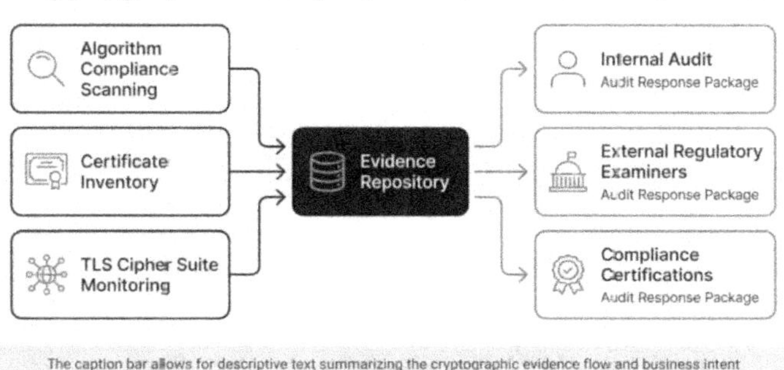

The caption bar allows for descriptive text summarizing the cryptographic evidence flow and business intent

9.4.1 **Building an Audit-Ready Cryptographic Control Framework**

An audit-ready cryptographic control framework maps each significant cryptographic security requirement to a defined control, a monitoring mechanism that generates evidence of control effectiveness, and a documented owner responsible for the control's ongoing operation. This framework structure is standard in information security compliance programs, and its application to cryptographic security makes the generation of compliance evidence systematic rather than reactive.

Controls in the framework should address the full scope of cryptographic security: algorithm selection requirements for new systems and updates to existing systems; certificate lifecycle management controls including issuance, renewal, and revocation procedures; key management controls including generation, storage, rotation, and destruction procedures; vendor cryptographic compliance requirements and assessment procedures; incident response procedures for algorithm breaks and deprecations; and monitoring controls that continuously validate algorithm compliance across the deployed environment.

The framework's value in an audit context is that it provides auditors with a structured view of the organization's control environment and a clear mapping between controls and evidence. An auditor examining cryptographic compliance can review the framework to understand what controls exist, then examine the evidence repository to confirm that the controls are operating as described. This is a fundamentally different — and more defensible — posture than attempting to reconstruct compliance evidence retrospectively in response to an audit request.

9.4.2 Evidence Collection and Continuous Monitoring for Algorithm Compliance

Continuous monitoring for algorithm compliance requires tooling that periodically scans the deployed environment for non-compliant algorithm usage and generates evidence records that can be used to demonstrate compliance at any point in time. This is distinct from point-in-time assessments conducted as part of an audit cycle — continuous monitoring provides the real-time visibility needed to detect compliance gaps as they emerge, rather than discovering them retrospectively.

The monitoring tooling should cover multiple layers: network protocol scanning to identify non-compliant cipher suites in TLS, SSH, and other protocol traffic; certificate inventory monitoring to identify non-compliant algorithms in issued certificates and to flag certificates approaching expiration; key management monitoring to identify non-compliant key types in managed key stores; and application scanning to identify non-compliant algorithm calls in deployed application code or container images.

Evidence records generated by continuous monitoring should be archived with sufficient retention to support audit cycles. A record that shows algorithm compliance status across the environment on each day of the past twelve months provides a much stronger compliance demonstration than a current point-in-time snapshot, because it shows that compliance was maintained continuously rather than just at the moment of audit. For heavily regulated organizations, the monitoring evidence archive may need to be preserved for the full regulatory retention period applicable to the systems being monitored.

Caption – Annual Algorithm Standards Review Cycle is a near infographith an executive business tone.

9.4.3 Regulatory Examination Preparedness in Federal and Healthcare Contexts

Federal agencies and healthcare organizations operate under specific regulatory frameworks — CNSS Policy 15, NIST SP 800-131A, HIPAA, HITECH — that will increasingly incorporate post-quantum cryptographic requirements as the standards landscape matures. Preparation for regulatory examinations in these contexts requires understanding not just what the current requirements say, but also what the examination process looks like and what types of evidence examiners typically request.

Federal examiners conducting cryptographic compliance assessments typically request: a current cryptographic inventory demonstrating that the organization knows what algorithms are deployed across its systems; evidence that approved algorithms are being used in accordance with the applicable standards; documentation of any deviations from the standards, including risk acceptance rationale and remediation timelines; and evidence of algorithm monitoring controls that detect non-compliant usage. Organizations that can produce these materials promptly and in a well-organized format are in a

241

much stronger position for examination than those that must reconstruct compliance evidence under examination pressure.

Healthcare organizations should anticipate that post-quantum cryptographic requirements will be incorporated into HIPAA enforcement actions and HITECH incentive programs over time, following the pattern of previous cryptographic updates. The organizations best positioned for this regulatory evolution are those that have already integrated post-quantum algorithm compliance into their HIPAA security rule compliance programs, rather than treating it as a separate obligation managed by a separate team.

9.5 Workforce Development for the Post-Quantum Enterprise

Cryptographic security depends on people who understand it well enough to make correct decisions at every level of the organization — from the developer choosing which library function to call, to the architect designing a new service, to the procurement officer evaluating a vendor's security claims, to the board member assessing whether the enterprise is taking the right level of risk. Building and sustaining that understanding across an organization is a workforce development problem, not a technology problem.

Diagram 9.3 – Role-Based Cryptographic Training Curriculum

■ Deep Coverage ▨ Standard Coverage ☐ Awareness Only

	Quantum Threat Basics	NIST Standards Overview	Implementation Practices	Compliance Obligations	Incident Response
Executive Leadership	Deep Coverage	Standard Coverage	Standard Coverage	Awareness Only	Awareness Only
Security Leadership	Deep Coverage	Deep Coverage	Standard Coverage	Awareness Only	Awareness Only
Security Engineering	Standard Coverage	Deep Coverage	Deep Coverage	Standard Coverage	Awareness Only
Application Developers	Awareness Only	Standard Coverage	Standard Coverage	Standard Coverage	Standard Coverage
IT Operations	Awareness Only	Awareness Only	Standard Coverage	Standard Coverage	Standard Coverage
Procurement and Legal	Awareness Only	Awareness Only	Awareness Only	Awareness Only	Awareness Only

■ Deep Coverage ▨ Standard Coverage ☐ Awareness Only

Tailored curriculum approach as the training curriculum is firm with a speak of cryptographic consistency.

9.5.1 Role-Based Training: What Different Functions Need to Know

A single cryptographic training curriculum cannot serve all organizational roles equally well. The depth, format, and content are not appropriate for a security engineer implementing ML-KEM in a TLS library; they are appropriate for an executive who needs to understand the business implications of a quantum threat but does not need to understand the key encapsulation mechanics. Role-based training acknowledges this reality and designs learning experiences calibrated to the decisions and behaviors each role needs to support.

Executive leadership and board members need a conceptual understanding of the quantum threat, the regulatory and competitive implications of inadequate cryptographic posture, and the resource and timeline commitments required to maintain quantum resilience. They need enough fluency in the topic to ask the right questions in a briefing — to distinguish between a roadmap that represents genuine progress and one that is managing perceptions — but they do not need implementation knowledge.

Security leadership — CISOs, security architects, and senior security engineers — need a working understanding of NIST standards, the performance and operational characteristics of standardized algorithms, the threat intelligence landscape surrounding quantum hardware advances, and the governance frameworks that sustain cryptographic compliance. This group is responsible for translating standards requirements into enterprise policy and for making algorithm selection decisions that balance security, performance, and operational constraints.

Application developers need practical, implementation-focused training: how to use the enterprise's approved cryptographic libraries correctly, how to recognize and avoid common cryptographic implementation errors, and how to identify when a task requires escalation to security engineering rather than independent implementation. This training should be heavily example-based and should use actual enterprise tooling and code patterns rather than abstract cryptographic concepts.

IT operations teams need training focused on configuration management and monitoring: how to configure post-quantum cipher suites in the protocols and platforms they manage, how to validate that post-quantum configurations are operating correctly, and how to recognize the monitoring signals that indicate a cryptographic issue requiring escalation. Operations teams do not need deep knowledge of algorithms, but they do need clear procedure-based guidance for the specific platforms they operate on.

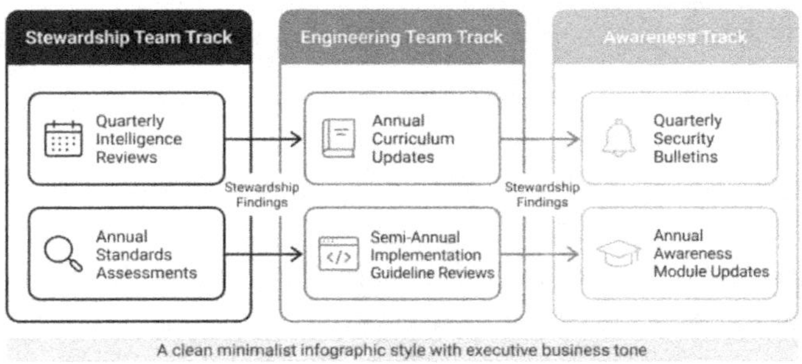

A clean minimalist infographic style with executive business tone

9.5.2 Building Internal Cryptographic Expertise Versus Sourcing It Externally

Every organization must make a deliberate decision about how much cryptographic expertise to develop internally versus source through external specialists, consultants, or managed security service providers. Neither extreme is sustainable: an organization that relies entirely on external expertise for all cryptographic decisions has no institutional memory and will be dependent on outside parties for every algorithm event, standards update, and vendor assessment. An organization that attempts to build deep cryptographic expertise for every function is investing in a level of specialization that most enterprises cannot staff or sustain.

The sustainable middle ground is to build sufficient internal expertise to own cryptographic governance and to make informed decisions about algorithm selection, vendor assessment, and compliance, while sourcing deep technical expertise — cryptographic implementation review, algorithm security analysis, and specialized standards interpretation — from external specialists who maintain that depth across many client environments and who track the research community continuously.

Internal expertise development should focus on the stewardship function: the team responsible for maintaining the asset inventory, managing the governance framework, and coordinating the enterprise response to algorithmic events. This team needs enough technical depth to evaluate vendor claims, interpret standards documents, and assess whether proposed implementations are sound. They do not need the research-level depth to evaluate novel algorithm constructions or to perform formal security proofs. That level of expertise, when needed, should be sourced externally from cryptographic research specialists.

9.5.3 Keeping Knowledge Current as Algorithms and Standards Continue to Evolve

The post-quantum cryptographic landscape will continue to evolve for years after the initial NIST standards are published. Additional algorithms will be standardized. Existing standardized algorithms may be deprecated or have their parameter sets updated in response to new cryptanalytic results. Protocol standards will be updated to incorporate new algorithm identifiers. Regulatory guidance will become more specific as the compliance frameworks catch up to the standards.

Keeping organizational knowledge current in this environment requires structured mechanisms, not ad-hoc awareness. The cryptographic stewardship team should have a defined process for monitoring relevant research and standards developments: participating in NIST standards forums, subscribing to academic preprint servers for cryptographic research, monitoring vendor security advisories and government cybersecurity agency publications, and regularly reviewing conference proceedings from major cryptographic research venues.

Annual knowledge reviews for the broader workforce — updated training modules reflecting current standards, algorithm advisories distributed through the security awareness program, and updated implementation guidelines in developer documentation — ensure that knowledge currency extends beyond the stewardship team to the engineers and operations staff who implement and operate cryptographic controls every day.

9.6 Strategic Threat Monitoring and Algorithmic Response Readiness

The ability to respond quickly when a post-quantum algorithm is broken, deprecated, or found to be vulnerable to a novel attack is one of the most important outcomes of sustained quantum resilience practice. Response speed depends on three things: early detection of the threatening development, pre-built response procedures that can be activated without starting from scratch, and an organizational structure with clear authority to make and execute response decisions quickly.

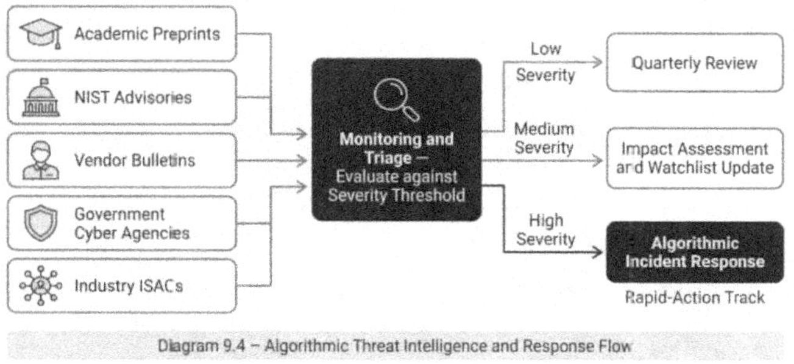

Diagram 9.4 – Algorithmic Threat Intelligence and Response Flow

9.6.1 Signals to Watch: Academic Breaks, Government Advisories, and Vendor Bulletins

The signals that indicate an algorithmic threat typically emerge through one of three channels before they generate regulatory or compliance requirements. Academic preprint publications are often the first place that significant cryptanalytic advances appear — a paper posted to a preprint server by a research team can signal a meaningful reduction in the security margin of a deployed algorithm weeks or months before it is formally published, peer-reviewed, and incorporated into government advisories or vendor bulletins. Organizations with a monitoring mechanism for cryptographic research preprints can detect these signals early and begin impact assessment before they become public news.

Government cybersecurity advisories — from NIST, CISA, NSA, and equivalent bodies in other jurisdictions — represent a more authoritative signal but often lag the academic publication timeline by weeks to months. When a government advisory recommends deprecating an algorithm or sets a sunset date for a specific algorithm, the enterprise's practical migration timeline begins with that advisory, not with the academic paper that first identified the vulnerability. Organizations that detect the academic signal early have more time to prepare before the government advisory creates compliance pressure.

Vendor bulletins and security advisories represent the third signal channel, typically focused on specific product vulnerabilities rather than algorithm-level weaknesses. A vendor advisory that a specific version of a TLS library implements ML-KEM with an error, or that a specific HSM firmware version has a side-channel vulnerability in its ML-DSA implementation, creates an immediate specific response requirement rather than a broad algorithmic threat. The enterprise must identify which systems use the affected product version and execute targeted remediation.

9.6.2 Maintaining a Rapid-Response Posture for Algorithm Deprecation Events

A rapid-response posture for algorithm deprecation requires pre-built capability, not just intent. When a significant algorithmic event occurs — a break, a deprecation recommendation, or a standards update — the enterprise should not have to build its response capability from scratch. The response infrastructure should already exist: the asset inventory that identifies affected systems, the runbooks that specify the remediation steps for each system type, the change management fast-track process that allows urgent security remediations to move more quickly than routine changes, and the escalation structure that ensures the right decision-makers are available within hours rather than days.

The organization should conduct periodic algorithmic incident response exercises — tabletop scenarios that simulate receiving a significant algorithm advisory and require the response team to walk through the identification, assessment, and remediation process under time pressure. These exercises reveal gaps in response capability before a real event requires it: missing runbooks, unclear authority for fast-track change-management approvals, incomplete coverage of asset inventories for specific system categories, or untested vendor-notification procedures.

Response readiness also requires maintaining the cryptographic asset inventory as a live operational tool, not an archived document. An inventory that accurately reflects the current state of cryptographic deployments across the enterprise is the foundation of any algorithmic response: it tells the team exactly which systems are affected by a specific algorithm deprecation, how many of them there are, and who is responsible for each one. An inventory that was accurate two years ago but has not been maintained since is worse than useless in a response scenario — it creates false confidence and wastes response time on systems that have already been migrated while missing newly deployed systems that have not.

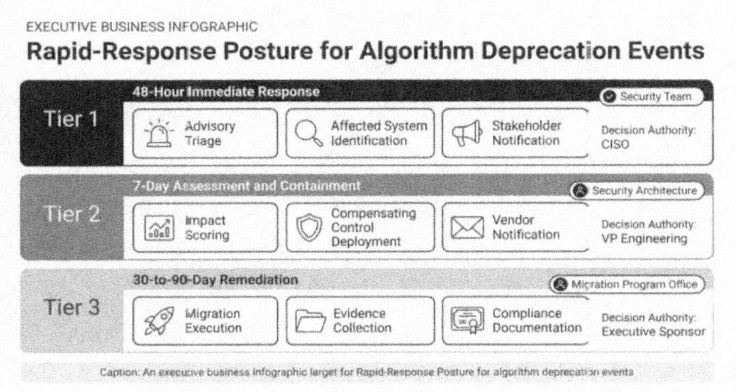

Caption: An executive business infographic larget for Rapid-Response Posture for algorithm deprecation events

9.6.3 Scenario Planning for a Faster-Than-Expected Quantum Threat Timeline

Most enterprise post-quantum migration timelines are calibrated to a consensus estimate of when a cryptographically relevant quantum computer might become available — a timeline that most published analyses currently place in the mid-to-late 2030s, with significant uncertainty in both directions. But an enterprise that has only planned for the consensus scenario is not resilient to the accelerated scenario: a faster-than-expected advance in quantum hardware or a breakthrough in error correction that significantly compresses the timeline.

Scenario planning for the accelerated timeline is not about predicting when the threat will materialize — no one can reliably do that. It is about identifying the conditions under which the enterprise's current migration timeline would be inadequate and determining what actions would accelerate the program if those conditions materialized. The key scenarios to plan for are: quantum hardware capabilities ahead of published consensus estimates; discovery of a significant vulnerability in a widely-deployed post-quantum algorithm before a replacement is available; and a geopolitical event that creates specific, credible intelligence about nation-state quantum capabilities that exceeds publicly available estimates.

For each scenario, the plan should identify which systems would be at elevated risk, which accelerated migration actions could be taken immediately, what resource requirements those actions would create, and what authority structure would govern the decision to activate the accelerated plan. Having this analysis pre-built allows the organization to move from scenario awareness to action in hours rather than weeks, the difference between proactive and reactive responses in a fast-moving threat environment.

9.7 The Institutional Posture That Outlasts Any Single Migration

The chapters preceding this one have described a structured migration program: inventory the cryptographic estate, design for algorithmic flexibility, phase the migration carefully, and govern the sustained posture. Executing that program well is necessary but not sufficient. The organizations that maintain quantum resilience over the long term are those that embed cryptographic lifecycle thinking into their institutional DNA — into how they procure technology, select vendors, manage partnerships, and allocate executive attention.

Institutional Cryptographic Lifecycle Thinking

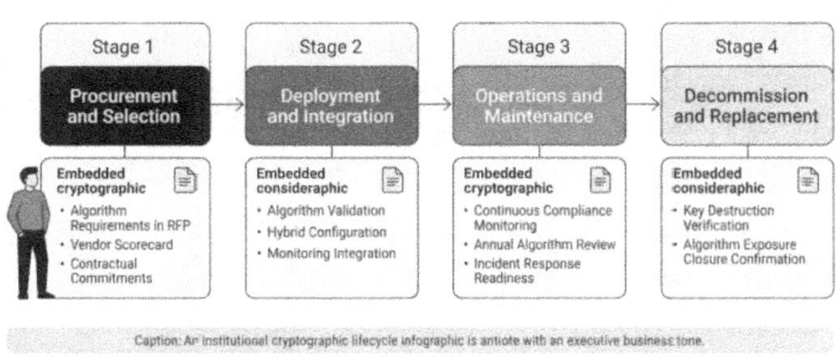

Caption: An institutional cryptographic lifecycle infographic is antiote with an executive business tone.

9.7.1 Embedding Cryptographic Lifecycle Thinking into Technology Procurement

Every significant technology procurement decision — a new application platform, a cloud service, a hardware device, a managed security service — has cryptographic implications that should be evaluated as part of the procurement process, not discovered after deployment. Technology that does not support current approved algorithms at the time of procurement will create a migration liability immediately upon deployment, regardless of how well the enterprise has managed the migration of its existing systems.

Embedding cryptographic lifecycle thinking into procurement means adding algorithm requirements to RFP templates and vendor evaluation criteria, assigning scoring weight to post-quantum algorithm support and vendor migration roadmap credibility, and requiring contractual commitments to algorithm update timelines as a condition of contract execution. It means training procurement staff to recognize algorithm-relevant questions and to escalate evaluations to the cryptographic stewardship team when vendor claims require technical assessment.

The payoff is that every new technology deployment starts from a post-quantum-ready baseline rather than adding to the migration debt that the enterprise must eventually work through. Over time, as the existing estate migrates and new procurements meet the post-quantum standard from day one, the enterprise's cryptographic technical debt shrinks to zero rather than growing indefinitely.

9.7.2 Post-Quantum Resilience as a Vendor and Partner Selection Criterion

The enterprise's quantum resilience posture is only as strong as the weakest cryptographic link in its value chain. Vendors and partners that have not achieved post-quantum resilience create exposure for the enterprise even after the enterprise completes its own migration: partner connections may revert to classical algorithms through negotiation fallback, vendor-managed systems may handle the enterprise's data without adequate cryptographic protection, and supply chain attacks may target less-prepared partners as an indirect path to the enterprise.

Post-quantum resilience as a vendor selection criterion means including it in the third-party risk assessment framework alongside conventional security criteria such as access control, incident response capabilities, and regulatory compliance. A vendor that scores well on conventional security criteria but has not begun post-quantum migration should be evaluated differently from one that has completed migration. The risk profile is different, and the enterprise's risk management decisions — whether to require remediation, accept residual risk, or choose a different vendor — should reflect that difference.

For strategic partnerships and critical suppliers, post-quantum resilience assessment should be conducted annually rather than only at initial vendor selection. A vendor that was adequately prepared at the time of selection may have fallen behind on its roadmap commitments. A vendor that was behind schedule at the initial selection may have caught up. Annual reassessment keeps the enterprise's picture of its third-party cryptographic risk current and identifies escalation needs before they become compliance failures.

9.7.3 **Sustaining Executive Attention Beyond the Initial Migration Cycle**

The most persistent organizational challenge in sustaining quantum resilience is maintaining executive attention after the initial migration program concludes. Migration programs naturally attract executive attention: they have visible timelines, measurable milestones, and a clear end state. Sustaining a governance practice does not have the same narrative pull. There is no completion event, no press release, and no board presentation that marks the "successful conclusion" of ongoing quantum resilience.

Sustaining executive attention requires connecting quantum resilience to the business and mission concerns that executives track continuously. When a regulatory body issues new guidance, that is an executive-relevant development that connects cryptographic posture to compliance risk. When a peer organization or competitor experiences a cryptographic incident, that is an executive-relevant development that connects cryptographic posture to reputational and operational risk. When a vendor changes its post-quantum roadmap in ways that affect the enterprise's posture, that is an executive-relevant development that connects cryptographic posture to vendor relationship management.

The cryptographic stewardship team should cultivate a practice of proactively identifying and communicating these connections, rather than waiting for executives to ask about cryptographic security. A brief executive advisory that explains why a recent academic publication about a post-quantum algorithm is relevant to the enterprise's posture and what the team is doing in response keeps executive attention engaged without requiring executives to monitor the cryptographic research landscape themselves.

Executive attention is ultimately sustained by demonstrated value. When the governance framework allows the organization to respond to an algorithm advisory in days rather than months, and when that response capability is communicated clearly to executive leadership, it reinforces the value of the institutional investment in sustained quantum resilience. The organizations that maintain that investment over the long term will be the ones whose cryptographic governance teams consistently demonstrate that the investment protects the enterprise from risks that would otherwise remain invisible until they materialize.

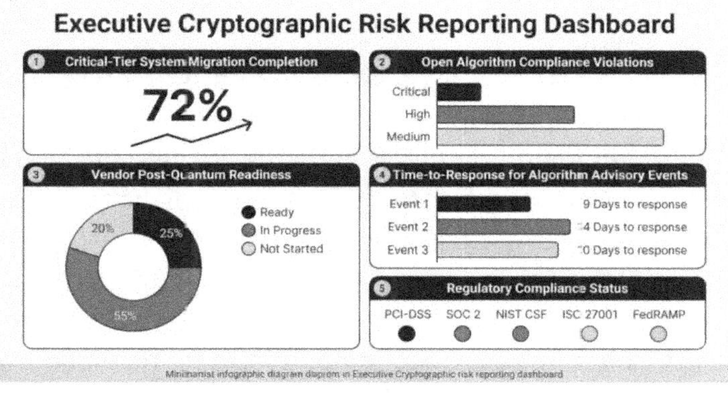

Minimalist infographic diagram diaprom in Executive Cryptographic risk reporting dashboard

9.8 Manager's Quantum Resilience Governance Checklist

Use this checklist to assess and build your organization's long-term quantum resilience governance posture:

- Assign named ownership to the cryptographic stewardship function. If this responsibility is distributed across existing roles rather than centralized, confirm that the owners are explicitly aware of their responsibilities and have the authority to execute them.
- Audit your policy architecture. Confirm that you have a cryptographic security policy, technical algorithm standards, and implementation guidelines, each with a named owner, a current review date, and a distribution channel that reaches the teams that need to follow them.
- Integrate quantum resilience into your GRC program as a defined risk domain. Confirm that cryptographic algorithm compliance is represented in the risk register, the control framework, and the internal audit scope.
- Confirm that executive and board reporting includes a cryptographic risk section at least quarterly, reporting on outcome metrics — not activity metrics — and identifying any escalations requiring leadership decision.
- Build the continuous monitoring infrastructure for algorithm compliance. Confirm coverage across at least three monitoring dimensions: network protocol scanning, certificate inventory, and key management monitoring.
- Develop role-based cryptographic training modules for at least four organizational roles: executive, security engineer, application developer, and IT operations. Schedule annual updates aligned to the algorithm standards review cycle.
- Establish a monitoring mechanism for cryptographic research publications and government advisories.

Define the internal process for triaging incoming signals and escalating those that require response action.

- Build and exercise an algorithmic incident response runbook. Conduct at least one tabletop exercise per year that simulates an algorithm deprecation advisory and requires the team to walk through the response process.
- Add post-quantum algorithm support requirements to all new technology procurement RFP templates and vendor evaluation scorecards. Require contractual commitments for algorithm update timelines in all significant technology contracts.
- Include post-quantum resilience status in the annual third-party risk assessments for all critical vendors. Escalate vendors whose migration timelines are inadequate for the enterprise's risk tolerance.
- Define the indicators and response thresholds for the accelerated quantum timeline scenario. Confirm that response authority and resource pre-authorization are in place so that the scenario can be activated within forty-eight hours of the trigger conditions being met.

9.9 Key Takeaways

Quantum resilience is not a project. It is an institutional capability that must be built, sustained, and continuously adapted as the threat landscape, the standards environment, and the technology estate evolve. Organizations that treat the post-quantum migration as a one-time program and then return to normal operations will find themselves rebuilding from scratch when the next algorithmic event demands a response.

The governance framework, the workforce competencies, the monitoring infrastructure, and the response readiness that sustain quantum resilience over the long term are not separate investments from the migration program — they are the program's completion. A migration that leaves behind a live inventory, a functioning governance structure, trained personnel, and exercised response procedures has accomplished its mission. A migration that leaves behind a static report and a decommissioned program office has only deferred the problem.

The long game in post-quantum security is not about predicting the exact timeline of the quantum threat. It is about building an organization capable of adapting to algorithmic change on any timeline — faster than expected, slower than expected, or in ways current models do not anticipate. That adaptability is what the governance, workforce, and monitoring infrastructure described in this chapter is designed to produce. It is the most durable investment the enterprise can make in its long-term cryptographic security posture.

10 The Migration That Cannot Wait for Certainty

You opened this book by confronting a mathematical reality that most enterprise leaders have not yet internalized: the cryptographic systems protecting your organization today rest on assumptions that a sufficiently powerful quantum computer will invalidate. The nine chapters that followed gave you the technical literacy to understand the threat, the operational framework to inventory your exposure, and the migration playbook to begin replacing vulnerable algorithms with the standards that have already been finalized and published.

The path you traced moved from algorithmic fundamentals through the NIST post-quantum selections, the cryptographic census that most organizations have never conducted, the hybrid deployment strategy that bridges the transition, and the crypto-agility architecture that ensures your enterprise is never locked into a single algorithmic generation again. You also confronted the governance and workforce dimensions that will determine whether the migration sustains momentum across the budget cycles and leadership changes it will inevitably span.

The temptation facing every leader who reaches this point is to wait. Wait for the quantum threat timeline to clarify. Wait for vendors to embed the new algorithms into existing products. Wait for a regulatory mandate that forces the issue. That wait is the single most dangerous decision available to you, because the harvest-now-decrypt-later threat is not a future risk. Adversaries are capturing your encrypted traffic today with the explicit expectation that quantum decryption will make it readable. Every month you delay the cryptographic inventory is another month of data that may be exposed retroactively, with no remediation possible afterward.

Begin with the census. Identify every certificate, key, and algorithm your organization depends on. Prioritize the systems where data confidentiality extends beyond the expected quantum timeline. Stand up the hybrid deployment for your highest-exposure communication channels. Assign a migration owner with cross-functional authority and a reporting line that reaches the executive team. These are not planning activities. They are the first operational steps of a transition that your organization will be executing for years.

The algorithms are ready. The standards are published. The only variable that remains within your control is whether your organization migrates on its own terms or under crisis conditions imposed by an adversary who moved first. This book gave you the framework. The decision to act is yours.

11 Resources Referenced

CISA. *Post-Quantum Cryptography Initiative.* Cybersecurity and Infrastructure Security Agency, 2024. (CH9).

European Union Agency for Cybersecurity (ENISA). *Post-Quantum Cryptography: Preparing for the Transition.* ENISA Report, 2024. (CH9)

Federal Register. "Announcing Issuance of Federal Information Processing Standards (FIPS) FIPS 203, FIPS 204, and FIPS 205." *Federal Register*, 14 Aug. 2024. (CH4).

Gidney, Craig, and Martin Ekerå. "How to Factor 2048-bit RSA Integers in 8 Hours Using 20 Million Noisy Qubits." *Proceedings of a Major Cryptography Conference*, 2025. (CH2)

Google Quantum AI. *Technical Report on Quantum Error Correction and Logical Qubit Scaling,* 2025. (CH2)

IEEE Standards Association. *Guidance on Cryptographic Agility and Post-Quantum Migration,* IEEE Std., 2025. (CH7)

IEEE Security & Privacy. Special Issue: *Post-Quantum Cryptography and Enterprise Migration,* IEEE, 2024. (CH3, CH8)

IETF. *RFC: Post-Quantum Key Exchange and Hybrid TLS Extensions,* Internet Engineering Task Force, 2025. (CH6)

IETF. *Draft: Best Practices for PQC Integration in Internet Protocols,* Internet Engineering Task Force, 2024. (CH6)

Jouguet, Paul, et al. "Practical Considerations for Post-Quantum TLS Deployments." *ACM Transactions on Privacy and Security*, 2025. (CH6)

Koblitz, Neal, and Alfred Menezes. "Post-Quantum Cryptography: A Survey of Lattice-Based and Hash-Based Approaches." *Journal of Cryptographic Engineering*, 2024. (CH3)

Lamport, Leslie. *Hash-Based Signatures: Theory and Practice.* Springer, 2024. (CH3)

Liu, Y., et al. "Hybrid Cryptography in Practice: Interoperability and Performance." *IEEE Transactions on Information Forensics and Security*, 2025. (CH6)

McKay, K., and S. Patel. "Cryptographic Inventory Techniques for Large Enterprises." *Journal of Cybersecurity Operations*, 2025. (CH5)

Microsoft Security. *Post-Quantum Migration Playbook for Enterprises,* Microsoft Security Research, 2025. (CH8)

National Institute of Standards and Technology (NIST). *Post-Quantum Cryptography | CSRC.* NIST Computer Security Resource Center, 2024. (CH4).

NIST. *FIPS 203: Module-Lattice-Based Key-Encapsulation Mechanism Standard,* 13 Aug. 2024. (CH4).

NIST. *FIPS 204: Module-Lattice-Based Digital Signature Standard,* 13 Aug. 2024. (CH4).

NIST. *FIPS 205: Stateless Hash-Based Digital Signature Standard,* 13 Aug. 2024. (CH4).

NIST. *IR 8547: Post-Quantum Migration Timeline and Guidance,* 2024. (CH4, CH8)

NSA. *Commercial National Security Algorithm (CNSA) Suite 2.0 Guidance and Transition Roadmap,* National Security Agency, 2024. (CH4, CH8)

Oxford Quantum Group. "Assessing Logical Qubit Roadmaps and Cryptographic Relevance." *Quantum Information Journal,* 2025. (CH2)

Pfleeger, Charles P., and Shari Lawrence Pfleeger. *Cryptographic Agility: Principles and Practice.* Wiley, 2025. (CH7)

Post-Quantum Cryptography Standardization Working Group. *Proceedings and Technical Reports,* 2024–2025. (CH3, CH4)

Quantum Economic Development Consortium (QED-C). *Enterprise Readiness for Post-Quantum Cryptography: Industry Survey,* QED-C Report, 2025. (CH5, CH9)

RSA Conference Proceedings. *Post-Quantum Migration Case Studies and Best Practices,* RSA Conference, 2024. (CH5, CH8)

Schneier, Bruce, et al. "Operationalizing Post-Quantum Cryptography in Large-Scale Systems." *Communications of the ACM,* 2025. (CH7, CH8)

Stanford Center for Research on Foundation Models. *White Paper: Long-Term Data Confidentiality and Harvest-Now, Decrypt-Later Threats,* 2024. (CH1, CH2)

☐ **US Department of Commerce.** *Guidance on Federal Adoption of Post-Quantum Standards,* 2024. (CH4)

US Office of Management and Budget (OMB). *Memorandum: Federal Agency Planning for Post-Quantum Cryptography,* 2025. (CH8, CH9)

Van Meter, Rodney, and K. Brown. "Hardware Security Modules and PQC Acceleration: An Evaluation." *Journal of Applied Cryptography,* 2025. (CH7)

World Economic Forum. *Quantum Computing and Cybersecurity: Strategic Implications for Enterprises,* WEF Report, 2024. (CH1, CH9)